M.O. 577 AIR MINISTRY

METEOROLOGICAL OFFICE

HANDBOOK OF METEOROLOGICAL INSTRUMENTS

PART II

INSTRUMENTS FOR UPPER AIR OBSERVATIONS

LONDON: HER MAJESTY'S STATIONERY OFFICE
1961

U.D.C. 551.508 (02)

Printed and published by
HER MAJESTY'S STATIONERY OFFICE

To be purchased from
York House, Kingsway, London W.C.2
423 Oxford Street, London W.1
13A Castle Street, Edinburgh 2
109 St. Mary Street, Cardiff
39 King Street, Manchester 2
50 Fairfax Street, Bristol 1
2 Edmund Street, Birmingham 3
80 Chichester Street, Belfast 1
or through any bookseller

Price £1 5*s*. 0*d*. net

Printed in England

CONTENTS

CHAPTER 5

SECONDARY RADAR FOR UPPER WIND MEASUREMENT

CHAPTER 6

RADIO DIRECTION-FINDERS FOR METEOROLOGICAL USE

CHAPTER 7

PRINCIPLES OF RADIO-SONDE DESIGN

CHAPTER 8

METEOROLOGICAL OFFICE RADIO-SONDES

CHAPTER 9

RADIO-SONDE CALIBRATION, BALLOONS AND ACCESSORIES

CHAPTER 10

INSTRUMENTS FOR METEOROLOGICAL OBSERVATIONS FROM AIRCRAFT

CHAPTER 11

METHODS OF MEASUREMENT IN CLOUD PHYSICS

LIST OF DIAGRAMS

LIST OF PLATES

HANDBOOK OF METEOROLOGICAL INSTRUMENTS

PREFACE

THE primary aim of this Handbook, as stated in the Introduction to Part I, is to give detailed information on the design, installation, operation and maintenance of the instruments used at Meteorological Office stations. Part I deals with the instruments used for surface observations, whilst Part II is concerned with the instruments used for observations and measurements in the free atmosphere. A few meteorological instruments, such as those used for measuring cloud height, could be classified either as surface or as upper air instruments. A distinction could be drawn between the two classes by defining upper air equipment as that which involves apparatus that is sent aloft. With two exceptions this criterion has been adopted in the Handbook, the exceptions being radar equipment for cloud and precipitation study and radio equipment for thunderstorm location.

Upper air measurements at meteorological stations are generally confined to levels up to 30 or 40 km. and the airborne instruments involved are carried by balloon or aircraft. Meteorological information at heights considerably above 40 km. is now being obtained, mainly on an experimental basis, by means of rockets and artificial satellites, but the scope of this book does not extend to the special instrumentation used in these vehicles. Information on this subject is to be found in recent publications on rocket techniques.

Much of the equipment now in use for upper air measurements is electronic in principle and is highly complex; this is particularly the case with radar applications. One of the difficulties in compiling a handbook on the subject is that of deciding how much technical detail should be included. It would require several volumes to provide all the detailed instructions that are necessary for the operation and maintenance of the more complex equipment. Moreover, it is now the general rule for technical manuals to be supplied with such equipment and it would be unnecessary duplication to include their contents in this Handbook. These manuals are primarily intended for the use of the specialist technicians whom most meteorological services now have on their staff. In preparing this Handbook, therefore, the aim has been to include sufficient information to give both the meteorologist and the specialist technician a comprehensive guide to the upper air instruments that are now available and to provide them with a book of reference on the subject. Where necessary the specialist is referred to the relevant technical manuals for detailed information on operation and maintenance. The subject necessitates the use of a number of technical terms that are familiar to the electronics expert but not necessarily to the meteorologist and a short glossary of these terms is included mainly for the benefit of the non-specialist reader. The definitions given in this glossary follow those recommended in *British Standards* 204[1]* and 1523[2].

---

* The superscript figures refer to the Bibliography on p. 193.

The requirements and general principles for the careful handling of instruments given in Chapter 1 of Part I apply with even more emphasis to upper air instruments. Most of these are more costly and more susceptible to damage than those used for surface observations. It is very important, therefore, that the utmost care should be used in their handling.

Although the methods and instruments used for upper air measurements are now almost as well established as those used for surface observations a considerable amount of further development is in progress and is, indeed, desirable. The brief account of the history of the subject given in Chapter 1 provides a background against which the present state of development can be viewed.

Information used in the preparation of Part II of this Handbook has come not only from the Instruments Division of the Meteorological Office but also from a number of sources outside the Office; these include instrument and radar manufacturers and research laboratories. War Office, Ministry of Supply and Admiralty technical manuals have also provided information on the military radar equipments that have been adapted for meteorological use. Where instruments of other countries have been described references to the sources of information are included in the Bibliography.

CHAPTER 1

HISTORICAL DEVELOPMENT

In this chapter it is proposed to give an outline of the history and progress of the development of the methods and instruments for investigating the upper air. Probably the first instrumental observation of the change in a property of the atmosphere with height was that made by the French writer and philosopher Blaise Pascal three centuries ago. A few years after Torricelli's invention of the barometer in 1643, Pascal found that the barometric reading decreased as he ascended the 52m. high Tour St. Jacques in Paris and this result was confirmed five years afterwards by Périer in experiments up to 1460m. on the Puy-de-Dôme. About one hundred years later another Frenchman, Pierre Bouguer, carried out observations on the slopes of mountains during a geodetic expedition to Peru and he determined the height of the freezing point in various latitudes. These early attempts, in which the instruments were carried to the levels of observation by the experimenters themselves naturally led to the consideration of some form of vehicle, such as a kite or a balloon, to carry the equipment into the free air. In the development of such aids there has been a constant striving to reach greater and greater heights. The development of instruments to be carried up into the air has been concentrated mostly on those for measuring pressure, temperature and humidity, wind and, to a lesser extent, on those for investigating the composition of the air and its electrical state.

1.1. KITE INVESTIGATIONS

The first serious attempt at scientific investigation of the free atmosphere was made in 1749 by Dr. Alexander Wilson of Glasgow, who used kites to raise thermometers to considerable heights in the air. Three years later Benjamin Franklin performed his famous kite experiments to prove the electrical nature of lightning. In 1822 kites carrying self-registering thermometers were used by the Rev. George Fisher and Captain Sir Edward Parry in Arctic voyages and in 1847 Birt[3] flew kites at Kew Observatory to measure changes of temperature and wind with height.

No systematic use was made of kites, however, until after 1885 when Dr. E. D. Archibald introduced steel piano wire as a means of reaching much greater heights. From then until the early years of the present century kites were the chief means of upper air investigation. From 1894 they were used regularly at Blue Hill Observatory, near Boston U.S.A., under the direction of L. Rotch; by Teisserenc de Bort, who founded the observatory at Trappes, near Paris, for the study of the upper air; by M. Rykatcheff at the Central Physical Observatory at Pavlovsk, near Moscow; by W. H. Dines and C. J. P. Cave in this country; and by H. Hergesell and R. Assmann in Germany where later the well known upper air observatory at Lindenberg was founded. The kites used were generally of the box type invented in 1893 by Hargrave of Sydney; one is shown in Plate I. The tension on the kite wire could exceed 100 kg. and a powerful winch was needed to control the flight.

The systematic kite ascents necessitated some form of continuous recording instrument and the first ascent with such an instrument was made at Blue Hill Observatory in 1894. The various designs which were evolved generally took the form of a baro-thermo-hygro-anemograph, and the names of Richard Frères,

C. F. Marvin, Bosch-Hergesell and W. H. Dines are associated with the best-known kite meteorographs. The instrument designed by Dines[4] is shown in Plate II. A shallow wooden frame carried pressure, temperature and humidity units controlling pens which marked on a circular chart rotated by a small clock mounted on the frame. The barometer consisted of a single aneroid capsule while the thermometer was a long spiral copper tube filled with alcohol and terminating at one end in a thin-walled box similar to an aneroid capsule. The free side of the latter operated the pen lever. A hair hygrometer was used. There was also an anemometer worked by the pressure of the wind on one or more light spheres (" ping-pong " balls) suspended by about 40 ft. of thread from a lever on the frame. The pull was balanced by a spiral spring, so arranged that the movement of the recording pen was proportional to the wind speed.

The excitement of kite flying as a means of meteorological investigation was vividly described by Cave[5] shortly before his death in 1950. By using several kites strung out along the wire it was possible to raise the instruments up to 6 km. or more, but there was always the risk of the wire breaking, sometimes with disastrous results. Such an occasion was that on which Teisserenc de Bort had eleven kites on 7 km. of wire trailing over Paris and the breakage resulted in stopping a steamer by the wire fouling its propeller and a train by jamming its connecting rods, and in cutting all communication with Brittany on the day when all France was anxiously awaiting the result of the famous Dreyfus case.

1.2. MANNED BALLOONS

Not many years after the first experiments of the Montgolfier brothers with hot-air balloons, the French physicist Charles invented the hydrogen balloon and in 1783 made an ascent in Paris, taking with him a barometer and a mercury thermometer. The first organized programme of meteorological observations in the free atmosphere was carried out by Dr. John Jeffries who ascended from London with the French aeronaut Blanchard with a barometer, thermometer (unventilated), hygrometer, electrometer and bottles of water for emptying and obtaining samples of air. The results were communicated to the Royal Society and included an average value for the rate of fall of temperature of 5°C. per km., which is reasonably consistent with more recent knowledge. The first trigonometrical height determination of a balloon flight was made a year later when Jeffries and Blanchard crossed the Channel in a balloon from Dover while French officers took readings of the balloon from Calais. Other balloon investigators in the early years of the nineteenth century included the well known French physicists, Jean Baptiste Biot and Louis Gay-Lussac, who in 1804 ascended to 7 km. and discovered that the proportion of water vapour in the air decreased with altitude but the chemical composition did not change.

A considerable advance was made in this country when the British Association initiated some upper air investigations in 1852. They started with some ascents from Vauxhall Gardens by Welsh[6] when he was Superintendent of Kew Observatory. An important point about Welsh's investigations is that he realized that he would not obtain very accurate air temperatures unless he reduced the effect of solar radiation on his thermometers. He did this by enclosing them in polished metal tubes through which air was forced by hand bellows, thus originating the aspirated apparatus later developed by Assmann. Welsh attained heights ranging from 4 to 7 km. and appears to have been the first observer to notice the occurrence of layers of air in which the fall of temperature was arrested.

The investigations under the auspices of the British Association were continued by James Glaisher who, with the aeronaut Coxwell, made 28 ascents and on one of them it was estimated that they reached the record height for that time (about 1865) of 11 km. This height, which must have been very near the tropopause, was not greatly exceeded by a manned balloon until seventy years later when Piccard and Cosyns reached 16 km. in 1932. A new record for a manned balloon flight was set up in 1957 by Captain J. W. Kittinger who ascended to 30 km. In comparing these records it should be remembered that Glaisher's ascent was made without any of the high-altitude safeguards now employed, such as oxygen apparatus, heated clothing and sealed cabins. Glaisher, in fact, lost consciousness for 13 min. In 1875 a high ascent was made by Gaston Tissandier with two companions who did not survive the effect of the great height. The equipment carried included, besides thermometers and barometers, an apparatus for estimating the amount of carbon dioxide by chemical absorption and a spectroscope for examining the water vapour line in the solar spectrum.

The danger to life at great elevations together with discrepancies between the results of manned balloon observations, some of which were obtained with poorly ventilated thermometers, led to the investigation of other methods, in particular the use of self-registering instruments on small unmanned balloons or " ballons-sondes " as they were first called.

1.3. SOUNDING BALLOONS AND METEOROGRAPHS

The first successful sounding balloon ascent with self-recording apparatus was that arranged in 1892 by the Frenchmen H. Hermite and G. Besançon who used an inextensible balloon made of gold-beater's skin of 4 m.^{3} capacity. This sounding reached 7·6 km. so to obtain a greater height a much larger balloon was constructed, which ascended to 16 km. Satisfactory temperature records were obtained up to 12 km. on a meteorograph constructed by Richard Frères of Paris but above that height the effect of solar radiation with the reduced rate of ascent became excessive. Satisfactory results at higher levels became practicable in 1901 when R. Assmann introduced extensible rubber balloons for carrying meteorographs. It was soon after Napier Shaw became Director of the Meteorological Office in 1900 that systematic upper air observations with sounding balloons began in this country. With much encouragement from Shaw, W. H. Dines devised the extraordinarily light and simple meteorograph which was first used in 1907 and became so well known.

1.3.1. Balloon meteorographs

The first balloon meteorographs were similar to those used in kite ascents. They generally incorporated an aneroid barometer and an alcohol-filled Bourdon tube thermometer. The large thermal inertia of the latter led Teisserenc de Bort to design a bimetallic thermometer using brass and steel soldered together. In order to reduce the weight of the equipment for use with his small rubber balloons Assmann designed an arrangement which dispensed with the rather heavy clock-work that had hitherto been used to drive the recording drum. In Assmann's instrument the recording chart, in the form of an endless band round two cylinders, was moved by the action of the aneroid barometer. Temperature and humidity were indicated transversely across the moving chart. The total weight of the instrument was 620 gm.

W. H. Dines's[7] balloon meteorograph also dispensed with a clock. It was a revolutionary design in other ways; in its original form as a baro-thermograph its weight was only 28 gm. without its aluminium case which weighed a further 35 gm. Plate III shows the instrument in its later form, incorporating a hygrograph. The frame DBCB′ is made from a single piece of nickel silver. The side of the aneroid capsule is rigidly attached at B and the upper end of the frame D forms a holder for a silvered recording plate. The lower end of the frame C acts as a spring tending to open the longer arms against the action of the aneroid which is connected by a thin spring E. Temperature is recorded by a bimetallic thermometer consisting of an invar rod HH′ rigidly fixed at B′ and to the bottom of a nickel silver strip MM, the top of which is connected to a steel multiplying lever EFP carrying a scriber point at P. The hair humidity element is a separate attachment secured to the frame by the screws SS′. It carries a spring-hinged arm with another scriber point. When the instrument is ascending the decreasing pressure causes the scribers to move across the silvered plate, while changes in temperature and humidity move them up or down the plate. The records scratched by the points are, therefore, functions of pressure. A light wire outer frame, attached only at the edges of the aneroid capsule, serves to carry a thin aluminium case and in use the whole instrument is protected from damage on falling by being suspended in a bamboo frame. The record is evaluated by means of a microscope with a magnification of about 80 diameters. A detailed description of the meteorograph, including its method of use and calibration procedure, has been given by L. H. G. Dines[8]. The instrument had the merit of being very light and accurate and it was cheap to make. Its lightness allowed the use of smaller balloons, yet the majority of soundings reached the stratosphere. The durability of the record was such that instruments recovered after remaining in the open for months or, as occurred in at least one case, after being fished up by a trawler from the bottom of the sea, gave readable results. The Dines meteorograph continued in regular use in this country for well over 30 years until it was superseded in 1939 by the radio-sonde.

Although not strictly a meteorograph, mention should be made of the ingenious balloon apparatus designed by Teisserenc de Bort for collecting samples of air at high altitudes. A barometric switch released a striker which broke off the tip of an exhausted glass tube. At a higher level the switch allowed an electric current to melt the remaining part of the tip and re-seal the tube. Samples of air up to 14 km. were thus obtained and subsequently analysed. A similar principle was used many years later in the sampling apparatus designed by L. H. G. Dines[9] and used in the well known work of Prof. Paneth on the chemical composition of the atmosphere.

A balloon meteorograph of special design, the alti-electrograph, was devised in 1937 for the work of Sir George Simpson and F. J. Scrase[10] on the distribution of electric charge in thunderstorms. Its principal feature was the use of a moving chart of pole-finding paper to record the sign of the point-discharge current produced by high electric fields in a wire hanging vertically from the apparatus. The instrument also gave a record of pressure and humidity.

1.4. PILOT-BALLOON HISTORY

The terms " pilot balloon " and " ballons d'essai " appear to have been introduced by navigators of manned balloons when they adopted the practice of releasing small paper balloons before an ascent in order to determine the probable direction of flight. The first attempt at a rough determination of wind flow above ground

W. H. DINES WITH AN INSTRUMENT-CARRYING KITE

Plate II

W. H. DINES' KITE METEOROGRAPH

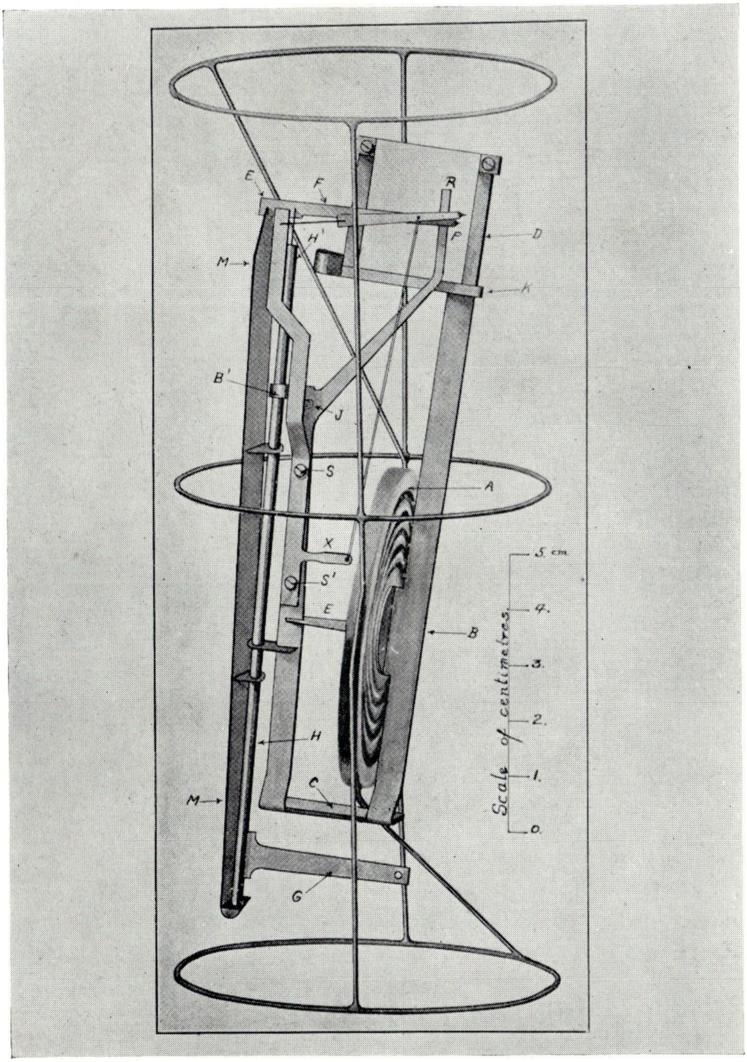

W. H. DINES' BALLOON METEOROGRAPH

DBCB′, frame
A, aneroid box
C, spring joint
D, plate holder
E, thin spring connecting piece
G, arm for calibration
K, spring

HH′, invar rod
MM, strip of nickel silver
EFP, steel multiplying lever
SS′, screws for securing hair humidity element
P, scriber point
R, arm
X, fixed arm

Plate IV

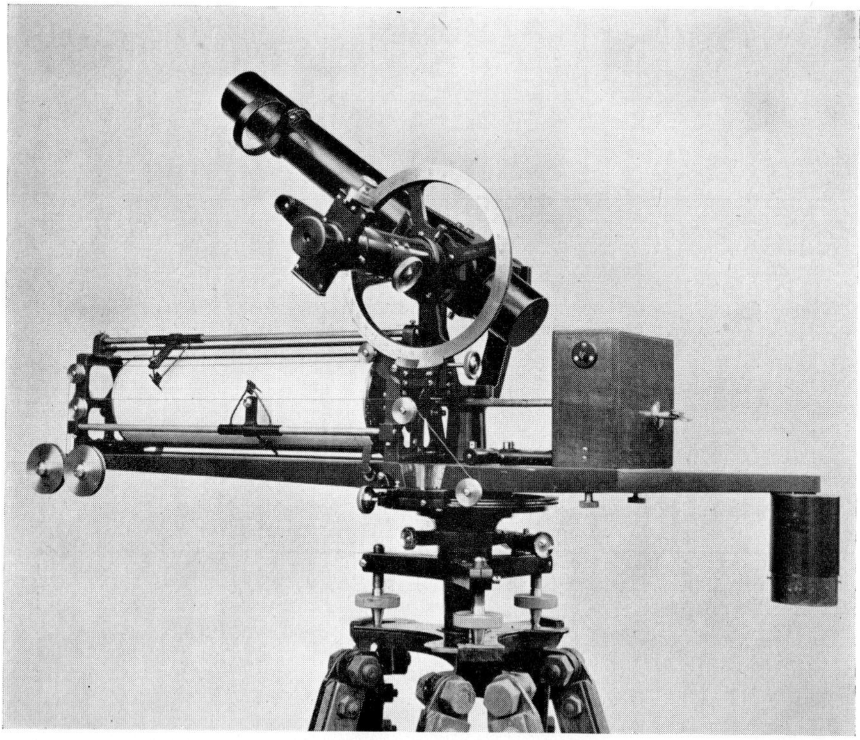

RECORDING PILOT-BALLOON THEODOLITE DESIGNED BY J. S. DINES

level is believed to have been made in 1809 by Thomas Forster who observed the drift of small free balloons filled with " inflammable gas ". It was not until 1873 or 1874, however, that the first instrumental measurements of the track of a large free balloon for the determination of the wind was made by Schreiber. In 1885 Schreiber again made such a measurement, using two surveying theodolites on a 5 km. base line, and reported the results in detail[11]. It is clear from his paper that he foresaw the usefulness of upper wind measurements in weather forecasting. In the following years similar measurements were made by other workers and in 1893 Kremser[12] tried also to determine the height of balloons by telescopic measurement of their apparent diameter. Up till this time such small balloons as were used were made of paper, but the rubber balloons introduced by Assmann in 1901 were found to have the advantage of practically constant rate of ascent. In 1903 H. Hergesell showed how the rate of ascent was related to the free lift and the weight of the balloon and with these advantages available pilot-balloon measurements began to be made systematically.

Before this, one or two attempts were made to adapt surveying theodolites for tracking sounding balloons. One example was the " dromograph " of Hermite and Besançon which registered automatically the azimuths and elevations of the balloon. These instruments were very much in the nature of improvizations and were not convenient for regular use. The first theodolite really suitable for pilot-balloon observation was that designed by de Quervain[13] in 1905. This instrument incorporated the right-angled telescope which enabled the axis of the eyepiece to remain horizontal no matter in which direction the telescope was pointed. The principle has been retained in modern designs, which have merely added refinements. Mention should be made, however, of a modification of de Quervain's instrument by J. S. Dines[14] in 1910, to make it self-recording. The recording attachment, which can be seen in Plate IV, was arranged so that changes in elevation and azimuth were recorded by two separate pens moving horizontally on a clockwork-driven drum.

The first systematic series of pilot-balloon observations in this country were those carried out from 1906 to 1910 by Cave[15]. He used both the single and double theodolite methods, first with a de Quervain theodolite made by Bosch of Strassburg, and later with a similar instrument made in this country by Cary Porter.

The introduction of radio and radar for tracking airborne objects naturally arrested the development of pilot-balloon methods, the use of which is limited by the presence of cloud and fog.

1.5. AEROPLANE METHODS

The introduction of sounding-balloon techniques did not entirely eliminate the need for the human observer to go aloft and when the aeroplane became a practicable vehicle it was soon adopted as a means of obtaining meteorological observations. In Germany the first aerological work with aeroplanes was undertaken in 1912 at Frankfurt-am-Main by Lt. von Hiddessen but it was during the First World War that regular flights for meteorological observations were commenced. They were undertaken not only by the belligerents but also by some of the neutral countries. In 1915 C. K. M. Douglas, then serving in the Royal Flying Corps, started making cloud and temperature observations (using ordinary thermometers suitably exposed) up to 3 km. in Northern France and in a paper[16] giving his results he strongly recommended the peacetime use of aircraft devoted solely to meteorological work.

Towards the end of the war Douglas took charge of a special meteorological flight which was equipped with photographic equipment that enabled him to obtain his well known cloud photographs. About that time wet- and dry-bulb psychrometers, using liquid-in-glass thermometers, were designed specially for fitting to the struts of the aircraft. Details of the early types of these have been given by Corless[17] who also described a baro-thermograph designed by Dobson in 1918 for use on aeroplanes. This instrument recorded temperature as a function of pressure as in the Dines's balloon meteorograph but was of such a size as not to require a microscope for reading the record. Other types of aircraft meteorographs such as the Jaumotte and the Bosch, more on the lines of the earlier kite meteorographs, i.e. with clockwork recording drums, were developed on the Continent.

The design of meteorological instruments for aircraft did not change much until the Second World War when higher speeds and high altitudes made the older types of instrument unsuitable. Electrical thermometers and psychrometers were introduced and also the frost-point hygrometer of Dobson, Brewer and Cwilong[18], which enabled measurements to be made of the minute amounts of water vapour present at high levels. Much of this development was associated with the investigations of the Meteorological Research Flight of the Meteorological Office.

1.6. APPLICATIONS OF RADIO AND RADAR

1.6.1. Development of telemetering

The development of methods of remote indication or recording of surface meteorological observations began in the second half of the nineteenth century and at the World's Fair in Philadelphia in 1876 the " universal telemeteorograph " on the chronometric principle built by the Dutch instrument-maker Olland in collaboration with von Baumhauer[21], in which transmitter and receiver were connected by telegraph wires, was displayed. Five years later, at an exhibition in Paris, a large number of designs for telemeteorographs were shown but apart from a few attempts to adapt these wired techniques to kite meteorographs no application was made to upper air work. Von Baumhauer had, in fact, suggested in 1874 such an application of his design to transmitting data from a kite or captive balloon and the instrument built to his specification by Olland would certainly have met the aerological requirements. It was not until 1917, however, that the first practical attempt at transmitting from a kite was made. This was done by F. Herath and M. Robitzsch, the transmission being through a single wire, namely the kite cable itself. With the aid of wireless transmission a new development of telemetering methods began about 1921 and was specifically directed towards aerological requirements. According to Wenstrom[20], Herath, in Germany, appears to have been the first to experiment, though not very successfully, with a crude spark transmitter attached to a balloon. Blair[21] of the United States Army Signal Corps reported having made similar experiments in 1923 in which signals were obtained to a height of 3·5 km. but the signals were not very good. It was not until the thermionic valve had reached a fairly advanced stage of development that there was any possibility of designing a satisfactory transmitter for upper air use.

1.6.2. Radio-sonde development

The work on the subject proceeded simultaneously in a number of countries but the credit for the first successful design of a balloon-borne radio transmitter is generally given to Idrac and Bureau[22] who, in France in March 1927, sent for the first time a small short-wave transmitter on a balloon into the stratosphere and received clear signals from the maximum height. In the development of a radio-sonde (the name suggested by Hergesell to correspond with the earlier term ballon-sonde) from this beginning it was natural that the chronometric method of telemetering, which had been satisfactorily demonstrated by Olland for surface transmission, should find favour. Molchanov[23] produced the first successful design in 1928 and this was improved upon in 1931 by Duckert[24, 25] who introduced a means of synchronizing the transmitter and receiver. By 1939 most of the larger meteorological services had developed their own designs of radio-sonde and descriptions were published by the International Aerological Commission[26]. The first attempt at radio-sonde design in this country was that of Thomas[27] in 1938 at the National Physical Laboratory. For various reasons the instrument was not adopted for use in the Meteorological Office. Instead, a new design, operating on the same general principle of audio-frequency modulation by variable inductance, was developed into what became known first as the Kew radio-sonde (the development having been undertaken at Kew Observatory by E. G. Dymond[28]) and later, when mass-production models were introduced, as the Meteorological Office radio-sonde.

1.6.3. Application of radio direction-finding

The application of radio direction-finding to the tracking of pilot balloons for upper-wind determination was reported by Blair and Lewis[21] to have been tried out in or soon after 1928 by the United States Army Signal Corps. The system, which measured azimuth only, involved measurements from two stations simultaneously. Tests of a specially designed balloon transmitter were made and it was found possible to track it with the direction-finders up to ranges of 15 km. In 1935 Corriez and Perlat[29] described direction-finding receivers which had been designed for the French Meteorological Service for upper wind determination.

The development of ultra-short-wave direction-finding equipment for the Meteorological Office was undertaken by the National Physical Laboratory and after trials in 1938 the equipment was adopted for routine measurements of upper wind. It continued to be very successfully used during the greater part of the Second World War. An account of the development of the direction-finders has been given by Smith-Rose and Hopkins[30]. Their use by the Meteorological Office has been described by Harrison[31].

The technique of thunderstorm location by the radio detection of atmospherics produced by lightning flashes may be said to have started in 1895 when the Russian physicist A. S. Popoff, using the coherer invented by E. Branly a few years before, constructed an untuned circuit for the reception of electromagnetic disturbances and demonstrated that they originated in lightning flashes. In 1906 G. Marconi suggested the possibility of relating the direction of atmospherics with the direction of distant storms from which they probably originated. It was in 1915 at Aldershot (South Farnborough), where the Meteorological Office had initiated research on the detection of thunderstorms by radio, that the first directional observations on atmospherics were made. This was done by Cave and Watson Watt[32] with an early type of radiogoniometer, the use of which had been suggested by R. Whiddington. The principle of the cathode ray oscilloscope was first applied to the direction-finding of atmospherics in 1926 by Watson Watt and Herd[33] and it continues to be the main feature of modern direction-finders for thunderstorm location.

1.6.4. Radio-theodolite development

The disadvantage of the two-station arrangement required by the azimuth direction-finding systems led to methods of measuring elevation being investigated so that a balloon could be tracked from a single station, as with the optical theodolite. Proposals for a method of measuring elevation as well as azimuth were made in 1937 both in the U.S.A. by Diamond, Hinman and Dunmore[34], and in the United Kingdom by Thomas[35]. The American proposal was to measure elevation by varying the height of the receiving aerial of the direction-finder until null response was obtained (due to the interaction of waves reflected at the ground with those received directly). The tests of this proposal were confined to ground experiments. Thomas's idea, which formed the subject of a British Patent, was to have an Adcock type of direction-finder with a pair of aerials spaced more than one wavelength apart. During the rotation of the equipment about a central vertical axis the received signal passes through a series of minimum intensity positions, two of which indicate the azimuth and the others the elevation of the transmitter. This method does not appear to have been pursued.

It was the military requirements of the Second World War which led to the successful development of the single station wind-direction-finder, i.e. the radio-theodolite. This was done by the Signal Corps Engineering Laboratories of the U.S.A. and the first equipment was known as radio set SCR–258. It operated on a frequency of 203 Mc/s. and was capable of tracking a balloon transmitter to a range of 150 miles. As this set was an adaptation of the first American radar equipment to be put into production it was large and cumbersome. A much smaller and more readily transportable set, the well known SCR–658, operating on a frequency of 400 Mc/s. was therefore specially developed. It was widely used in field operations. After the war it was superseded by rawin set AN/GMD–1, operating on the much higher frequency of 1,680 Mc/s. and incorporating such refinements as automatic following. An account of these developments has been given by Kirkman and LeBedda[36].

1.6.5. The application of radar

The suitability of the anti-aircraft gun-laying type of primary radar equipment for the measurement of upper winds from a single station was recognized early in the Second World War. The measurement of the range of a balloon-borne reflector as well as its azimuth and elevation proved to be very advantageous from the point of view of accuracy. About 1943 the Meteorological Office adopted the Army Radar Set A.A. No. 3 Mark 2, more familiarly known as the GL 3, for this purpose. With various modifications to improve its performance it has continued to provide one of the most accurate means of determining upper winds.

The use of secondary radar (with automatic re-transmission instead of reflection) for this purpose seems to have been first adopted during the war by Germany, where the equipment known as the " Fledermaus " was developed. In this system a transponder carried by the balloon received an audio-frequency modulated signal from a ground station on a carrier frequency of 300 Mc/s. and re-transmitted it on a frequency of 30 Mc/s. The slant range was found from the phase difference between the audio-frequency note transmitted from the ground and the note received at the ground from the transponder. The advantage of secondary radar in enabling balloons to be tracked to longer ranges than was possible with primary radar led to the design, soon after the war, of the radar-theodolite system for the Meteorological Office by Jones, Hooper and Alder[37]. The main difference between this system and the " Fledermaus " is that both the outgoing and returning transmissions of the radar-theodolite system are in the form of pulses, the time delay between which is measured to ascertain the range of the transponder. Additionally the return pulse is on a centimetric wavelength thereby affording a relatively high angular resolution of the transponder's position.

1.7. ROCKETS

During recent years improvements in balloons have enabled heights of about 45 km. to be reached. This, as explained in Section 9.4.4. however, is the maximum height likely to be attained with balloons, so attention has been turned to rockets as a direct means of investigating conditions at greater heights. Although rockets are believed to have been used as weapons as far back as the thirteenth century, it was not until the third decade of the present century that serious consideration was given to the design of a rocket for sounding the upper air. The first practical design of a rocket to carry sounding equipment was probably the WAC Corporal which was flight-tested in the U.S.A. in 1945. It was intended to carry 11·5 kg. of instruments to at least 30 km. but it was not used for upper air research because supplies of the German military rocket, the V2, became available for this purpose in 1946. The V2, which could readily reach 100 km. with a heavy pay-load, was adapted for various upper air investigations.

The main meteorological quantities which can be measured or investigated by rockets are pressure, temperature, density, ozone content, solar radiation and wind. In general the instruments used differ greatly from those designed for balloon sounding and in some cases are not yet out of the experimental stages. The first observations with registering apparatus carried into the ozonosphere by a V2 were obtained in 1947. In that year also the first Aerobee rocket, which had been under development in the U.S.A., was launched and reached a height of 60 km. Two years later Viking rockets reached more than twice this height with heavy loads of instruments. More recently a less expensive type of rocket called a Rockoon has been developed. This is a small cordite-fuelled rocket with a jet-assisted take-off unit for launching from a plastic balloon at heights of up to 25 km. It has reached 100 km. During and since the International Geophysical Year (1957–58) large numbers of rockets have been used by the U.S.S.R. for meteorological measurements up to about 100 km.

Rockets for upper air research have been developed in at least three other countries within the past few years. In France a rocket known as Veronique is designed to carry 50 kg. to 100 km. and there is a British one known as Skylark which can carry a similar pay-load to 300 km. A most important advance was achieved in the U.S.S.R. in 1957 by the launching of the first artificial earth satellite, Sputnik I, from a rocket at a height of about 800 km. The possibilities opened up by these new vehicles are vast and will undoubtedly lead to a considerable widening of the field of meteorology.

1.8. CHRONOLOGICAL SUMMARY

The principal dates in the history of the development of aerological methods are summarised in the following list:

1643 Pascal discovered the decrease in barometric reading with height up a tower.

1749 Dr. Alexander Wilson used kites to raise thermometers in the air.

1752 Benjamin Franklin made kite experiments on lightning.
1784 Dr. John Jeffries made the first manned-balloon ascent for meteorological observations.
1804 Gay-Lussac and Biot ascended in a balloon to 7 km. to investigate the composition of the air.
1809 Forster observed the movement of small balloons filled with " inflammable gas ".
1822 Rev. George Fisher and Capt. Sir Edward Parry raised self-recording thermometers on kites in the Arctic.
1847 Birt flew kites at Kew Observatory to obtain changes of temperature and wind with height.
1852 Welsh of Kew Observatory made four manned-balloon ascents to measure pressure, temperature and humidity.
1862 Glaisher reached an estimated height of 11 km. in a balloon.
1873 Schreiber made the first instrumental measurements of the track of a balloon.
1874 The telemeteorograph was designed by von Baumhauer and Olland.
1885 Dr. E. D. Archibald introduced the use of steel piano wire for kites.
1892 The first sounding-balloon ascent with self-recording apparatus was made by Hermite and Besançon.
c. 1893 Hargrave invented the box-kite.
1894 The first kite ascent with continuous recording apparatus was made at Blue Hill Observatory.
1895 Popoff detected lightning flashes by their electromagnetic disturbances.
1901 Assmann introduced extensible rubber balloons.
1905 De Quervain designed the balloon theodolite.
1906 Cave commenced systematic pilot-balloon ascents in England.
1907 The first ascent was made with the Dines meteorograph.
1912 Aeroplanes were used for meteorological observations in Germany by Lt. von Hiddessen.
1915 Douglas started meteorological observations in British aeroplanes.
1915 Cave and Watson Watt made the first directional observations on atmospherics.
1918 A psychrometer for use in aeroplanes was designed in the Meteorological Office.
1918 Dobson designed a baro-thermograph for use in aeroplanes.
1926 Watson Watt and Herd applied the cathode ray tube to atmospherics direction-finding.
1927 Idrac and Bureau first obtained radio signals from a balloon transmitter in the stratosphere.
1928 Molchanov produced a design for the first radio-sonde.
1928 Blair and Lewis first applied radio direction-finding to the tracking of pilot balloons.
1937 Sir George Simpson and F. J. Scrase first used alti-electrographs on balloons to investigate thunderstorm electricity.
c. 1940 The first radio-theodolite (SCR–658) was developed in the U.S.A.
1942 Dobson designed the frost-point hygrometer for aircraft observations.
c. 1943 Radar was used for the measurement of upper wind.
c. 1946 Rockets were first used for upper air research.
1957 The first artificial earth satellite was launched by the U.S.S.R.

CHAPTER 2

PRINCIPLES OF UPPER WIND MEASUREMENT

2.1. GENERAL CONSIDERATIONS

2.1.1. Basic measurements

Upper winds are most commonly measured by observing the movement of an airborne object, such as a balloon, ascending at a more or less uniform rate. In radio and radar methods, as opposed to optical methods of tracking the movement, the balloon must carry a radio transmitter or a radar reflector. Other methods that are occasionally used include the observation of the motion of clouds, shell bursts, smoke puffs and, more recently, of " window ", the electrically conducting material devised for screening military radar targets. There are also methods, using navigational principles, of measuring upper winds from aircraft.

The basic information that is required is the plan position of the airborne object at known time intervals, and its height. This information can be obtained by measuring the azimuth of the object from the point of observation and two of the co-ordinates, angle of elevation, slant range and height. If two observing stations at the ends of a base line of known length are employed the measurements can be reduced to one co-ordinate, other than height, at both stations and one of the other co-ordinates at one of the stations.

Upper wind speed is generally reported in knots and the direction from which the wind is blowing in degrees from true north, in a clockwise direction.

2.1.2. Effect of fluctuations

The wind at a given level in the atmosphere is subject to fluctuations due to turbulence. In the absence of such fluctuations a balloon would travel along a smooth track and its movements over equal time intervals would vary smoothly. The effect of the turbulent fluctuations is to produce a random distribution of vector displacement of the actual positions of the balloon from the hypothetical mean positions on the smooth track. Errors of observation produce displacements of the same type, but differently distributed and of different magnitude. It is impossible to distinguish between these two sources of scatter in the observations, but for routine measurements of the upper wind it is not necessary to do so. Some form of smoothing process is applied in order to remove the scatter and arrive at a mean wind over an interval of a few minutes, which, since the balloon is ascending, is equivalent to a height interval or layer thickness of a few thousand feet. It is important to know over what interval to take the mean wind. This involves, in addition to some knowledge of the instrumental errors, a decision as to the thickness of the layer, or duration of time, over which fluctuations are significant.

2.2. CLASSIFICATION OF METHODS

The methods available for the measurement of upper wind by tracking balloons can be divided into optical (pilot-balloon) theodolite, radio direction-finding and primary and secondary radar techniques.

The use of the pilot-balloon theodolite is, of course, limited by poor visibility and by cloud. The method allows accurate measurements of azimuth and elevation but if height is based on the assumption of a uniform rate of ascent it is liable to large errors. These can best be avoided by using two stations at a known distance apart. The errors can also be reduced in the earlier stages of a pilot-balloon ascent by measuring the angle subtended by a tail attached to the balloon.

The need for obtaining upper wind information in cloudy weather and in bad visibility has led to the widespread use of radio and radar methods of tracking the airborne object. In the early application of radio direction-finding as described, for example, by Harrison[31], the balloon carried a transmitter, the plan position of which was determined from the azimuth measured simultaneously from two or more stations separated by distances of the order of 20 miles. The measurement of elevation as well as azimuth by what is now known as a radio-theodolite enables the plan position of the transmitter to be determined from a single station, thus avoiding the need for what was usually an expensive system of intercommunication in the two-station method. Another system, employed during the Second World War, depended on the measurement of slant range instead of elevation. With any of these systems some means of determining the height of the balloon is necessary if the assumption of constant rate of ascent is to be avoided. This is usually effected by turning the transmitter into a radio-sonde measuring pressure and temperature.

In the application of radar to upper wind measurement the airborne object is tracked by the radar beam from which azimuth, elevation and also the slant range are obtained. This method therefore has the advantage of providing satisfactory determinations of the height as well as the plan position of the balloon. In primary radar systems the slant range is derived from the time of passage of a signal from a ground transmitter to and from a suitable reflector carried by the balloon. In secondary radar systems the reflector is replaced by a transponder which re-transmits the signal received from the ground transmitter.

2.3. EFFECT OF OBSERVATION ERRORS ON WIND MEASUREMENTS

2.3.1. Errors of observation

Reports of upper winds usually refer to measurements of the mean wind in a layer a few thousand feet in depth, during the passage of which two or three observations of the balloon may be taken. The errors in the mean wind arise partly from errors in the observed plan positions of the balloon and partly from errors in the height (or pressure) of the layer. The vector difference between two plan positions is less dependent on the actual magnitudes of the errors of the observations of the two positions than on the change in these errors between successive readings. In other words, a random error which introduces scatter into the graph of the observations is worse than a systematic error which shifts the graph as a whole but has little effect on its shape. Height errors, which can result in the measured wind being attributed to the wrong level, are not usually important unless the heights are derived from an assumed rate of ascent. Heights measured by radar methods are affected more by systematic errors, such as those arising from incorrect levelling of the equipment, than by random errors.

The errors in the angular readings of a radar set on land can be measured by comparison with optical theodolite observations. On ocean weather ships sextant readings provide a standard of comparison for the radar elevations. Alternatively the same balloon can be tracked by two instruments of the same type near together and from the differences between these readings their errors of observation can be obtained.

2.3.2. Effect of errors of observation on vector wind error

The effect of the errors of the measurements of azimuth, elevation, slant range and height on the vector wind error depends on the method of observation. In deriving the relationships between the errors for the various methods the following symbols are here used:

$\delta\mathbf{V}$ = vector error in computed wind $\mathbf{V}$

δS and δD = random errors in wind speed S and direction D

δl and δr = random errors in horizontal range l and slant range r

δe and δa = random errors in elevation e and azimuth a

δh = random error in height h

$Q = \bar{\mathbf{V}}/\bar{v}$ = ratio of mean vector wind and mean rate of ascent up to the height h

T = time to reach height h

t = time interval between consecutive observations

δv = random variation in mean rate of ascent between consecutive soundings

B = length of two-station base line

The mean square vector error in the computed wind is given by

$$(\delta\mathbf{V})^2 = (\delta S)^2 + \mathbf{V}^2 (\delta D)^2 .$$

This can be expressed in terms of plan-position errors thus:

$$(\delta\mathbf{V})^2 = \frac{2}{t^2}[(\delta l)^2 + l^2 (\delta a)^2] .$$

As already mentioned the vector wind may also be in error if, as a result of an error in the height, it is attributed to the wrong level. The random error in the mean height of a layer is $(\delta h\sqrt{2})/2$ and the resulting error $\delta\mathbf{V}_h$ in the vector wind will depend on the vertical gradient $\delta\mathbf{V}/\delta h$ of the wind, thus

$$(\delta\mathbf{V}_h)^2 = \tfrac{1}{2}(\delta\mathbf{V}/\delta h)^2 (\delta h)^2 .$$

From the above formulae, together with the substitutions

$$l = r \cos e ,$$

$$\cot e = (r^2 - h^2)^{\frac{1}{2}}/h = \bar{\mathbf{V}}/\bar{v} \equiv Q ,$$

the relationships between the random errors of measurement and the mean square vector wind and height errors may be derived. The vector errors vary primarily with range and elevation but they are more usefully expressed in terms of the height and the ratio Q of the mean wind to the mean rate of ascent of the balloon. These relationships for the various methods of observation are as follows:

(i) *Radar*, measuring slant range r, azimuth a, and elevation e

$$(\delta\mathbf{V})^2 = \frac{2}{t^2}\left\{\frac{(\delta r)^2 Q^2}{Q^2+1} + (\delta e)^2 h^2 + (\delta a)^2 h^2 Q^2\right\}$$

$$(\delta h)^2 = \frac{(\delta r)^2}{Q^2+1} + (\delta e)^2 h^2 Q^2$$

(ii) *Single theodolite, optical or radio,* measuring azimuth a and elevation e, combined with radio-sonde measurement of pressure–height.

$$(\delta \mathbf{V})^2 = \frac{2}{t^2}\left\{(\delta h^2)\, Q^2 + (\delta e)^2\, h^2\, (Q^2+1)^2 + (\delta a)^2\, h^2\, Q^2\right\}$$

$(\delta h)^2$ depends on radio-sonde errors

(iii) *Double theodolite or direction-finder,* measuring azimuth a (and elevation e if height is determined); a_1 is the azimuth, with respect to the base line, measured from the station from which the balloon is released

$$(\delta \mathbf{V})^2 = \frac{2\,(\delta a)^2\,(B^2 - 2Bh\,Q\cos a_1 + 2h^2\,Q^2)\,(B^2 - 2Bh\,Q\cos a_1 + h^2\,Q^2)}{B^2\,t^2\,\sin^2 a_1}$$

$$(\delta h)^2 = \left\{\frac{h^4\,Q^4 + (B^2 + h^2\,Q^2)^2}{B^2\,Q^2}\right\}(\delta a)^2 + \frac{h^2\,(Q^2+1)^2\,(\delta e)^2}{Q^2}$$

(iv) *Single theodolite, optical or radio,* measuring azimuth a and elevation e, but with height h derived from an assumed rate of ascent,

$$(\delta \mathbf{V})^2 = (\delta v)^2\, Q^2 + \frac{2\, v^2\, T^2}{t^2}\left\{(\delta e)^2\,(Q^2+1)^2 + (\delta a)^2\, Q^2\right\},$$

$$(\delta h)^2 = (\delta v)^2\, t^2/2$$

2.3.3. Relative accuracy of methods

The foregoing formulae enable the vector wind and the height accuracy obtainable by the main methods of measurement to be assessed when the random observational errors of the primary measurements are known. To illustrate this we may take the following somewhat idealized values of the quantities involved:

$\delta r = 20$ m. $\qquad t = 1$ min.

$\delta e = 0{\cdot}1^\circ$ $\qquad B = 10$ km.

$\delta a = 0{\cdot}1^\circ$ $\qquad a_1 = 90^\circ$

$\delta h =$ height error equivalent to a pressure error of 1mb., for method (ii).

$\delta v =$ 1m./sec., (2 kt.), for method (iv).

With these values the vector error of wind as a function of height and of the ratio Q of the mean wind to the rate of ascent is given in Table I.

TABLE I—VECTOR ERROR OF WIND AS A FUNCTION OF HEIGHT AND OF RATIO Q OF MEAN WIND TO RATE OF ASCENT

Q	5				10				15				20			
	a	*b*	*c*	*d*	*a*	*b*	*c*	*d*	*a*	*b*	*c*	*d*	*a*	*b*	*c*	*d*
	Height (km.) — knots															
1	1	1	1	2	1	2	2	3	2	3	3	3	2	6	5	4
2	1	2	2	4	2	5	5	6	3	8	13	7	4	13	18	9
3	1	4	3	7	3	9	11	10	4	14	23	13	5	22	40	17
4	2	7	5	10	3	14	18	16	5	22	40	22	6	33	71	28
5	2	11	7	14	4	22	28	23	6	33	61	32	8	48	110	42

a—radar.
b—optical or radio-theodolite and radio-sonde height.
c—double theodolite or direction-finder.
d—optical or radio-theodolite and assumed rate of ascent.

It is evident from Table I that the radar method, which gives a direct measurement of slant range, is much more accurate than the other methods which do not measure this quantity. For example, with a rate of ascent of 5 m./sec. (10 kt.) and a mean wind of 30 kt., i.e. $Q = 3$, the vector error at 10 km. with the radar method is 3 kt. but with the other methods it is about 10 kt. It should be pointed out, however, that while the errors assumed in the table are representative of those of a good radar set the azimuth errors of radio direction-finders might well exceed 0·1° and accordingly give larger vector wind errors than those given in the table. On the other hand good optical theodolites should be capable of 0·02° angular accuracy, in which case the vector errors would be one fifth of those given in the table for the methods using these instruments. The errors for the double-theodolite or direction-finder method increase rapidly when the balloon goes beyond a range (Qh) much greater than the length of the base line; they also increase roughly in proportion to the cosecant of the azimuth with respect to the direction of the base line. It would appear from Table I that the advantage of the greater accuracy of the double-theodolite method compared with that of the single theodolite at the shorter ranges is lost when the range exceeds about 30 km., i.e. three times the length assumed for the base line. It must be remembered, however, that in the expression for the vector wind error of the single theodolite, although random errors or variations in the rate of ascent affect only the first term, errors in the assumed rate of ascent, due for example to incorrect inflation, affect the other terms and cause systematic errors in the vector wind (and in the derived heights). These systematic errors may exceed those due to random variations and the chief advantage of the double-theodolite method is to eliminate them.

Using the same values of the random observational errors the corresponding root mean square errors in height are given in Table II.

TABLE II—ROOT MEAN SQUARE HEIGHT ERROR AS A FUNCTION OF HEIGHT AND OF RATIO Q OF MEAN WIND TO RATE OF ASCENT

Q	Height (km.) 5				10				15				20			
	a	*b*	*c*	*d*	*a*	*b*	*c*	*d*	*a*	*b*	*c*	*d*	*a*	*b*	*c*	*d*
	metres															
1	17	14	27	43	22	25	51	43	29	54	85	43	37	111	130	43
2	19	14	29	43	35	25	69	43	52	54	136	43	69	111	216	43
3	26	14	34	43	51	25	92	43	77	54	187	43	102	111	314	43
4	34	14	45	43	68	25	123	43	102	54	244	43	136	111	415	43
5	43	14	55	43	85	25	163	43	128	54	310	43	170	111	515	43

a—radar.
b—optical or radio-theodolite and radio-sonde height.
c—double theodolite or direction-finder.
d—optical or radio-theodolite and assumed rate of ascent.

It will be seen that the random height errors with a radar having 20 m. range error and 0·1° angular errors are roughly the same as those derived from a radio-sonde with a pressure error of 1 mb. The relatively small error at the greater ranges, arising from a random variation of 1 m./sec. in rate of ascent in the single-theodolite method, would generally be exceeded by a systematic error arising from the assumption of an incorrect rate of ascent and the effect of this on the values of the vector wind and the height would be proportional to the systematic error. This might well be as much as 10 per cent. and thus give an error of 1 km. at a height of 10 km.

The effect of the height error on the vector wind may be illustrated by taking the figure of 128 m. given in Table II for the radar case with $Q = 5$ and $h = 15$ km. With an unusually high vertical gradient of vector wind of 0·04 kt./m. the resulting error in knots would be $0 \cdot 04 \times 128/\sqrt{2}$, i.e. 3·6 kt. This may be compared with 6 kt. given in Table I for the corresponding error in vector wind arising from the plan-position errors. The total error in this case would be $\sqrt{(6^2 + 3 \cdot 6^2)}$, i.e. 7 kt. In fact, the effect of the random height error in the radar method is in almost all cases negligible, but the effect of systematic errors in height, arising, for example, from incorrect levelling of the radar or theodolite or from the assumption of an incorrect rate of ascent is not necessarily negligible and in single-theodolite ascents may be large.

CHAPTER 3

EQUIPMENT FOR PILOT-BALLOON OBSERVATIONS

3.1. GENERAL REQUIREMENTS

3.1.1. Description of method

In the pilot-balloon method of measuring upper winds the balloon carries little or no load and so can be quite small. The balloon is inflated with hydrogen (or helium) so as to rise at a convenient rate. The path of the balloon is then tracked by measuring, by means of a theodolite, its azimuth and elevation at successive intervals of time. Being light for its size the balloon takes up with negligible lag the direction and speed of the air in which it is floating at any particular instant. In addition to azimuth and elevation a knowledge of the height of the balloon at the times of the measurements is required. In the simplest but least accurate form of observation the height is obtained by assuming a constant rate of ascent of the balloon. Some improvement on this procedure can be obtained, at short ranges, by attaching a tail to the balloon and measuring the angle subtended by it. Clearly, this measurement cannot be made with reasonable accuracy when the balloon is more than a few miles away.

When high accuracy is required it is necessary to avoid all assumptions about the rate of ascent of the balloon. This can be done by measuring azimuth and elevation from two points at the end of a carefully surveyed base line.

To obtain the wind speed and direction from the observations the plan positions of the balloon at the successive times of the observations are computed by trigonometry and the winds are derived from the displacement of the plan position in the known time intervals. Full details of the operational procedure for pilot-balloon measurements are given in a Meteorological Office publication[38].

3.1.2. Equipment required

The following items constitute the equipment needed for pilot-balloon observations:

Pilot balloons

(i) Balloons specified by weight, ranging from 20 to 100 gm. and by colour, blue, white or red.

(ii) Balloon filler for determining free lift.

Hydrogen

(i) Hydrogen cylinders, with lever key for the main valve, or hydrogen generator. (Information about hydrogen equipment and the precautions to be taken in using it are given in Section 10.2.6. of Part I).

(ii) Fine-adjustment valve for controlling the rate of flow of the gas.

(iii) Pressure gauge for measuring pressure in the hydrogen cylinder.

(iv) Pressure-gauge adapter to enable the gauge and the fine-adjustment valve to be used simultaneously.

(v) Rubber tubing to connect the balloon filler to the fine-adjustment valve.

Theodolite

(i) Pilot-balloon type.

(ii) Theodolite tripod or wallhead.

(iii) Light filters (fitted to eyepiece or object glass).

Accessories

(i) Stop-watch.

(ii) Ordnance survey map of district, or prismatic compass.

(iii) Computing aid, such as special slide-rule or graphical device.

(iv) Thread, coloured paper, aluminium wire, and release device for tail ascents.

(v) Lantern and candles, or electric lamp and batteries for night ascents.

(vi) Thread and elastic for attaching lantern to balloon.

(vii) Watch oil and vaseline, camel hair brush and chamois leather.

3.2. PILOT BALLOONS

3.2.1. Description

Pilot balloons are of the extensible rubber type and are spherical when inflated. They should be made of high quality suitably compounded rubber latex, cured or vulcanized by a heat process. Synthetic rubber, such as neoprene, may also be used instead of the natural product. Other materials such as gutta-percha, paper, varnished silk and plastics have been used but they are not as satisfactory as rubber. The finished balloons should be free from foreign matter, pin-holes or other defects and must be homogeneous and of uniform thickness. The necks should not be less than 2 in. in length when the balloons are inflated. Some details of modern manufacturing methods are given in Section 9.4.1. on radio-sonde balloons.

3.2.2. Care and handling

Balloons should be kept in the dark in a place that is not subject to extremes of temperature. Arrangements for the replenishment of supplies should be such that the period of storage before use is reduced to a minimum. Before the balloon is inflated any loose French chalk should be shaken out. The balloon should then be warmed and rubbed well between the hands. This reduces the risk of a burst during inflation and helps to make the balloon assume a regular shape. Exposure to strong sunshine before an ascent is very injurious and should be carefully guarded against.

3.2.3. Sizes and colours

For upper wind measurement four sizes of pilot balloons are generally used. They were formerly specified by their nominal circumference when fully inflated but are now specified by their nominal weights. Table III gives details of those in use in the Meteorological Office.

TABLE III—SIZES OF PILOT BALLOONS

Stores Ref. No. Met.	Nominal weight	Nominal circumference inflated
	gm.	*in.*
57	20	70
58	30	90
429	80	200
601	100	250

Actual weights may differ from the nominal weights by amounts up to about 20 per cent. The 20 gm. balloons are used for relatively low ascents (when it is expected that the height reached will be under 10,000 ft.). The 30 gm. balloons are used for ascents when the cloud base is at a height of 10,000 ft. or more, for ascents when conditions are especially favourable for high soundings, and for ascents when tails or lanterns are used. Both the 20 gm. and 30 gm. balloons are inflated to give a rate of ascent of 500 ft./min. For special purposes when higher rates of ascent are required the 80 gm. and 100 gm. balloons can be used. The former is normally filled to rise at 700 ft./min. and the latter 1,000 ft./min.

Of the three colours normally available for pilot balloons white (or untinted) is usually found to be best in clear cloudless weather, as the sun's rays are reflected by the balloon which consequently appears as a bright point in the sky. White balloons are unsuitable at low elevations, i.e. with high winds, owing to hazy conditions near the horizon. Blue balloons should be used when the sky is overcast, while red balloons are usually more suitable on days of broken cloud, when the background varies between blue sky and white cloud. In general, red is the most useful colour and blue the least useful.

The visibility of the image of the balloon in the theodolite is sometimes improved by the use of one or other of the coloured light filters which are normally provided for fitting to the eyepiece of the telescope. These filters absorb the short-wave scattered light from haze and mist and improve the contrast between the balloon and its background. Different colours, such as dark red, light red, yellow and green, are used according to the state of the atmosphere.

3.2.4. Free lift and rate of ascent

A detailed discussion of the factors affecting the rate of ascent of rubber balloons is given in Section 9.4.3. dealing with radio-sonde balloons. For most pilot-balloon work, the following simplified formula is adequate:

$$V = q\,L^{\frac{1}{2}}/(L + W)^{\frac{1}{3}}$$

where V is the rate of ascent, L the free lift, W the weight of the balloon and its attachments and q is a numerical factor depending mainly on the drag coefficient. When the rate of ascent is expressed in ft./min., q is 275 for 20 gm. and 30 gm. balloons and 310 for 80 gm. balloons. Larger balloons are subject to variations in the drag coefficient and it is found, for example, that for the 100 gm. size q varies from 320 when the balloon is filled to rise at 700 ft./min. to 375 when it is filled for 1000 ft./min.

3.2.5. Balloon fillers

The amount of inflation necessary to produce the required free lift and rate of ascent can be determined by using either a filling nozzle of the required weight or one which forms the arm of a balance on which the lift can be weighed. The latter is less convenient unless it is desired to allow for differences between actual and nominal weights of balloons, which is hardly necessary in routine work.

The balloon fillers in use in the Meteorological Office are essentially hollow brass or light alloy nozzles shaped to accommodate balloon necks of different diameters and containing a one-way ball valve. With the fillers are provided brass weights which can be screwed on as required. Plate V shows a filler of this type. The valve end is connected to the hydrogen supply by means of a length of rubber tubing and the other end of the filler, with the appropriate weight attached, goes into the neck of the balloon. Hydrogen is then admitted until the filler is just lifted off the ground. When the rubber tubing is disconnected the balloon should float with the filler attached. Any excess hydrogen can be let out by pressing the stem of the filler valve. With the filler removed the balloon should then ascend at the required rate. Three varieties of fillers for pilot balloons are available for use at meteorological stations. They are listed in Table IV. The rates of ascent corresponding with the free lifts obtainable with the appropriate fillers and weights are given in Table V.

TABLE IV—METEOROLOGICAL OFFICE PILOT-BALLOON FILLERS

Mark No.	Stores Ref. No. Met.	Filler weight	Additional weights	Application
		gm.	*gm.*	
8*	1776	20	(A) 41·5 (B) 51·5 (C) 60·5 (D) 10	20 gm. and 30 gm. balloons with or without tails and lanterns
5	537	71·5	(A) 166·5	30 gm. and 80 gm. balloons
7	624	230	(A) 151 (B) 292 (C) 475	100 gm. balloon with or without electric lamp and battery

* The Mark 8 filler is similar to a Mark 4 which it supersedes.

TABLE V—SELECTION OF FILLERS AND WEIGHTS FOR REQUIRED RATES OF ASCENT

Balloon and attachments	Filler Mark No.	Weight to be added	Free lift	Rate of ascent
			gm.	*ft./min.*
10 gm.	8	None	20	400
20 gm.	8	A	61·5	500
20 gm. + 30 ft. tail	8	B + tail	66·5	500
20 gm. + 120 ft. tail or lantern ..	8	B + tail, or B + D	81·5	500
30 gm.	8	B	71·5	500
30 gm. + 120 ft. tail or lantern ..	8	C + tail, or C + D	90·5	500
30 gm.	5	None	71·5	500
80 gm.	5	A	238	700
100 gm.	7	None	230	700
100 gm. + 85 gm. electric torch ..	7	A	381	700
100 gm.	7	B	522	1000
100 gm. + 85 gm. electric torch ..	7	C	705	1000

If weight D is not being used with the Mark 8 filler when a tail or lantern is required then the tail or lantern *must* be attached to the filler to obtain the correct adjustment of the free lift. When weights A or C are used with the Mark 7 filler the torch *must not* be attached while the balloon is being filled.

3.3. PILOT-BALLOON THEODOLITES

3.3.1. General features

Fundamentally the theodolite consists of a telescope mounted so that it can rotate both in a vertical and a horizontal plane. The balloon theodolite differs from the surveying type in having an optical system so designed as to permit the axis of the eyepiece to remain horizontal no matter in which direction the telescope is pointed. This is arranged by using a right-angled prism or a pentagonal prism. The latter is preferable since a slight displacement of it does not affect the perpendicularity of the two parts of the optical axis. The magnification of the telescope is usually between 20 and 25 diameters and the field of view not less than 2°. Cross-wires or a graticule are fitted to the eyepiece which has a focussing adjustment. The mounting of the telescope is such that it can be turned rapidly by hand and slowly by gearing on azimuth and elevation circles. These are graduated in degrees or less and are provided with verniers or micrometer hand wheels to allow reading to 0·01°. The base of the theodolite is designed for fitting on to a standard tripod or other support, with adjustments to allow accurate levelling and orientation. To facilitate the tracking of a rapidly moving balloon open sights are provided and, in some instruments, an auxiliary telescope with a wide field of view, e.g. not less than 8°. A means of illuminating the scales is usually provided for use at night.

The fundamental principle underlying the construction of pilot-balloon theodolites has remained unchanged since the first one was designed by de Quervain in 1905. Later designs have been in the nature of refinements to achieve compactness, weather-proofing and ease of operation. The Meteorological Office pattern balloon theodolite is typical of the instruments now in general use, but before describing it in detail brief mention will be made of two special classes of theodolite, viz., one that incorporates a recording mechanism and the other designed for use in a ship.

Recording theodolites.—Probably the earliest form of recording theodolite was that devised in 1910 by Dines[14], and referred to in Section 1.4. and Plate IV. It was an adaptation of the earlier type of balloon theodolite and recorded azimuth and elevation by linear movements of pens and a clockwork drum, the pen carriages being controlled by the telescope circles through chains and pulleys. In some later instruments of this class the indications of the azimuth and elevation circles are recorded photographically or by means of a printing device. No great advantage is achieved since the subsequent computation involved is the same as for eye readings. Some advantage is gained, however, by those theodolites which provide a graphical record of the plan position of the balloon on a small polar co-ordinate chart immediately below the body of the theodolite. The record is semi-automatic, being made by the observer actuating a device which puts a point on the chart. The small size of the chart limits the accuracy with which the wind speed and direction can be derived from the record of the plan position of the balloon. It also limits the range to which the balloon's path can be recorded to about 60,000 ft. when the rate of ascent is 600 ft./min. Details of instruments of this type are given by Schoute[39] and Scultetus[40].

Marine theodolite.—The motion of a ship makes the problem of pilot-balloon observation a very difficult one. Theodolites have been adapted for shipboard use by mounting them on the top of a weighted rod which hangs from a gimbal or Cardan spring suspension. Even with the provision of damping, however, it is impracticable to use an ordinary theodolite so mounted except in a relatively steady ship. It is better to use the sextant principle for measuring the angle of elevation of the balloon, and theodolites specially designed for marine use incorporate an instrument of this type together with a horizontal circle for azimuth measurement (usually relative to the ship's head). If the sextant is of the bubble type the observations are not dependent on a natural horizon being visible.

3.3.2. Meteorological Office pattern theodolite

Description.—The Mark 1 design of the Meteorological Office pattern theodolite is described in detail by Cranna[41]. It is illustrated in Plate VI while the latest model, the Mark 4 (Stores Ref., Met. 661), is shown in Plate VII. It will be seen that the azimuth and elevation are read from two circles, E and F, arranged horizontally, one above the other. The upper one is the elevation circle and it is operated by the movement of the telescope axis through bevel gears which are kept in close engagement by means of a spring H. In the current model the circles are graduated in half-degrees.

The circles, and the corresponding telescope movements, are operated by tangent screws and worm-wheels through spindles carrying drums which for each complete rotation turn the telescope through one degree. The drums are divided into tenths of a degree. Both circles are read through the same window on the glass of which a reference line is engraved. This window and the drums are on the same side of the instrument as the eyepiece. The drum for the azimuth movement is on the observer's right, with the lever J immediately above for disengaging the worm-spindle; the elevation drum is on the left, with its lever below. There is a second window on the opposite side of the instrument; scales and drums may be read from this side if desired. The direction of rotation of the elevation drum of the Mark 2 design is opposite to that of the Mark 1 and these designs also differ in the method of operation of the elevation circle. In the Mark 1 the elevation circle and worm-wheel are in one piece and the elevation drum moves the telescope in elevation through both the worm and bevel gears. In the Mark 2 the elevation worm-wheel is mounted on the telescope axis so that the worm moves the telescope in elevation directly; the bevel gear now only transmits this movement to the elevation circle.

In addition to the main telescope, which views through a right-angle by means of a pentagonal prism and has a magnification of 20 diameters and a field of view of 2°, there is an auxiliary sighting telescope A with one quarter of the magnifying power and a correspondingly wide field of view. The same eyepiece serves for both telescopes and it is fitted with a graticule. The auxiliary telescope can be brought into operation by a pivoted mirror actuated by a lever B close to the auxiliary object glass. In the Mark 1 and Mark 2 instruments the focus of both telescopes is adjusted by a knurled ring C behind the eyepiece. The latter, in all models, is focussed on the graticule by screwing the eye lens D in and out. Mark 1 and Mark 2 theodolites also differ in the position of the open sights. On the earlier model these are on the outer side of the main telescope tube; on Mark 2 they are on the top of the mirror housing of the auxiliary telescope.

Illumination is provided for the circles and the graticule. For the circles a small electric bulb is carried in a screened holder just above the front scale window. To illuminate the graticule a similar bulb is carried in a holder K, so arranged that the beam of light is thrown on to a small mirror in the centre of the telescope tube, whence it is reflected on to the prism and so on to the graticule which appears dark against an illuminated field. The degree of illumination can be varied by turning the knurled disc L. Switches controlling these lights can be seen in Plate VII; the dry batteries are housed in the upper part of the body of the instrument.

Some Mark 1 and all Mark 2 theodolites are provided with a hook for holding a stop-watch and with a reading lens. The watch hook is so located that the watch is illuminated by the scale lamp. The reading lens is arranged so that the circles may be read to 0·1° with the minimum movement of the head, and without reference to the drums, a procedure preferred by some observers.

The Mark 4 theodolite differs from the Mark 2 in the following respects:

(i) The focus of the main telescope is fixed

(ii) The optical system includes a larger object glass and pentagonal prism, which result in a brighter and more uniform field and improved definition

(iii) The battery box is detachable

(iv) A set of colour filters is included as a standard fitting

(v) A rubber eyeshield is provided

(vi) The rim of the rotating eye-lens is graduated so that an observer may set the eyepiece to his own focus with the minimum delay

(vii) More durable materials are used for the worms and worm-pivots to reduce wear

(viii) The mirror of the auxiliary telescope is mounted so that it is more readily adjustable.

The Mark 3 instrument is transitional between the Marks 2 and 4 and embodies all the changes except (ii), (iv), (v) and (vi).

A tripod (Stores Ref., Met. 591) is provided for use with the theodolite. It has a threaded head on which the sub-base of the instrument fits. When the tripod is not in use the threads are protected by a screw-cap with a knurled edge. A fitting for the cap is provided on one of the legs to prevent its loss while the tripod is in use. The cap is also provided with a hexagon key which enables it to be used for tightening or loosening bolts on the tripod.

3.3.3. Setting up the theodolite

The necessity of exercising great care in handling a theodolite cannot be too strongly stressed. A fall usually strains the parts so badly that the instrument cannot be repaired with completely satisfactory results.

The site where the theodolite is used should be carefully chosen. It should have a clear unrestricted view, especially in the direction down the prevailing wind. It should be as far from any obstructions, such as buildings or trees, as possible but if there is a small building, such as a hut for storing instruments, balloons and hydrogen, it is preferable that the observing station should be on the lee side of it. The ground on which the station is sited should be firm. If the theodolite is set up on unpaved ground, this is apt to get very muddy in wet weather. It is better, therefore, that gravel should be laid. For a permanent station a pillar fitted with a wallhead may be used instead of a tripod to support the theodolite.

The tripod should be set up firmly on the ground. The protecting cap should be unscrewed from the tripod head and the hexagonal key should be used to tighten the hexagon-headed screw on the head, taking particular care not to strain this screw; the head is rigid when there is no play between the six balls and their supports at the tops of the three legs and the screw should not be tightened more than is just sufficient to remove any play.

The theodolite should now be carefully removed from its case and screwed on to the tripod head. It will be found easier to turn the instrument initially in the direction of unscrewing until the threads have engaged, and then to screw up until the theodolite is securely held. The use of excessive force in screwing up should be avoided or the threads may be jammed. The cap of the object glass should now be removed and placed in the box. By means of the tripod legs, the instrument can be brought to a convenient height, with its base approximately level.

Levelling the theodolite.—The azimuth clamping screw (N, Plate VI) should be fastened and the azimuth tangent screw released. The theodolite should now be levelled, first in a direction parallel to the line joining two of the levelling screws and then (by turning the third screw) in a direction perpendicular to that line. The following points should be noted:

(i) When making the first of these adjustments the screws should be turned in opposite directions by about the same amount, so as to avoid tilting the theodolite in the direction perpendicular to the line joining these screws.

(ii) The spirit-level is not always exactly perpendicular to the axis of rotation and the level setting is not therefore always obtained by making the bubble come to the centre of the glass tube. The only reliable criterion is that the bubble shall not move in the tube when the instrument is turned through 180°. It is not necessary to satisfy this criterion with a high degree of precision, as on the Meteorological Office pattern theodolites even a whole scale division of movement of the bubble does not give rise to an error in the reading of elevation greater than 0·01°. The method of checking the correctness of the spirit-level mounting is given in Section 3.3.4.

(iii) Any alteration of level made in one direction as a rule slightly disturbs the level in a direction at right-angles, consequently successive slight adjustments must be made to each setting alternately before both can be made correct.

Focussing.—With the Mark 3 and Mark 4 theodolites it is only the eyepiece that needs focussing. This is done by directing the telescope towards the sky so that the field of view is illuminated and then rotating the eyepiece until the graticule divisions are seen distinctly on the field of view. If now the telescope is directed to a distant object, both the object and the graticule divisions should be sharply defined in the field of view and there should be no relative motion of the two (parallax) when the observer's head is moved slightly from left to right.

In the case of Mark 1 and Mark 2 theodolites, which have focussing adjustment on the telescope, this adjustment must be used, after the eyepiece has been focussed, in order to bring the image of the balloon into the plane of the graticule. In that position the image will be sharply defined but accuracy of adjustment requires in addition that there shall be no movement of the image of the balloon relative to the graticule as the eye is moved from side to side. If the image moves the same way as the eye it is further away from the observer than the graticule and the graticule must be screwed in, and *vice versa*. If the image of the balloon moves on the graticule (without the eye being moved) merely as a result of altering the focussing of the telescope it is best to focus on a fixed distant object before beginning the observations, even though this means that the balloon is slightly out of focus during the first minute or two.

Orientation of horizontal circle.—As a sighting object some distant point should be selected whose true bearing from the observer is known (either from a large-scale ordnance map or by compass measurement). After fastening the azimuth clamping screw (N, Plate VI) the telescope should be rotated in azimuth until the reading on the horizontal circle is equal to the bearing of the selected point, the final setting being made with the azimuth tangent screw (I, Plate VI) which is then left engaged but must not be touched during the following procedure. The azimuth clamping screw N should be released and also the elevation tangent screw; it will then be possible to rotate the telescope in azimuth without altering the azimuth reading. The telescope should be directed towards the sighting object, using the open sights. Then, after re-engaging the elevation tangent screw and the azimuth clamping screw N, the object should be brought to the centre of the graticule by means of the elevation tangent screw and the screw P. The azimuth circle will then be orientated to indicate bearings from true north and the screws P and N should not be touched again.

Auxiliary telescope.—The small telescope should be used only as a finder and not for measurements. It may not be possible to adjust it to agree sufficiently accurately with the main telescope either in orientation or in focus. (This can easily be seen by taking readings on a fixed object with both telescopes.) Moreover, the focal length does not bear any simple ratio to that of the main telescope, so measurements of a tail with it are useless.

3.3.4. Care and adjustment of theodolites

General.—The theodolite is an expensive precision instrument and it is important that it should receive great care; nearly all defects in theodolites are caused by careless treatment by observers who fail to realize that they are using delicate instruments. Great care should be taken not to strain the parts of a theodolite by the use of excessive force. Parts such as locking nuts, screws, etc., should be screwed home firmly but gently, by hand or by a tommy bar, but they must not be forced in any way or the threads may be stripped.

If the instrument is being transported from one place to another, care should be taken to avoid jarring it, to prevent any strain being thrown on the axes. All clamps should be tightened before transporting the instrument and it should not be carried fixed to its tripod. All nuts and bolts of the tripod should be kept tight up against the woodwork and the metal shoes should be kept sharp.

When the instrument is in use and it is desired to turn the telescope by hand it is important that the tangent screw slow motions should be completely disengaged from their respective circles, otherwise the grating of the teeth of the worm gearing will cause undue wear.

A theodolite should not be left exposed out-of-doors when it is not in use. A waterproof cover may be placed over it if it is to be used again within a few hours. It should be replaced in its box when observations have been completed. When putting a Meteorological Office pattern theodolite in its box the azimuth tangent screw should be left in engagement, the elevation tangent screw put out of engagement and the telescope turned to an elevation of about 90°. The azimuth clamping screw should then be loosened and the base of the theodolite turned until the spirit-level is above the orientation adjusting screw. With the first finger of the right hand round a levelling screw and the top of the theodolite held in the left hand the instrument should be carefully lifted into the box so that the eyepiece is in the marked position and the telescope fits firmly down horizontally with its end on the baize-covered bracket. The elevation tangent screw should then be put into engagement, the purpose of leaving the tangent screws engaged being to take the tension off the springs.

Periodical examination.—The theodolite should be examined and tested periodically in order that any faults that develop may be quickly recognized and corrected. The theodolite has the advantage, in this respect, of being a self-checking instrument, it being possible to ascertain the correctness of adjustment of the instrument from its own readings without the use of auxiliary apparatus. Instructions for the examination of theodolites are set out below, together with the methods of correcting the various faults. If the necessary skill required to correct faults revealed by the tests is not available at a Meteorological Office station the theodolite should be returned to the Instrument branch. With a theodolite in regular use items (i) to (vii) should receive attention every week and the remaining items every two months.

(i) *General condition.*—Cleanliness of the instrument is as important as careful handling. Dust in the bearings increases the wear and tends to clog the parts, while dust on the lenses impairs the field of view. Dust should be removed with a camel hair brush; no abrasive such as emery should ever be used in cleaning the instrument. During the cleaning of the theodolite and tripod an examination of the general condition should be made to see that no screws are missing and that no parts have developed undue stiffness or slackness.

(ii) *Lenses.*—The surfaces of the lenses and the circle windows should be dusted with a camel hair brush and smears should be removed with a clean chamois leather or soft tissue paper. Before replacing the lenses any particles of fluff left by the cleaning material should be removed with the brush. Compound lenses should not be dismantled for cleaning.

(iii) *Graticule.*—Any dirt will show up as specks in the field of view. To clean the graticule it should first be removed by screwing it on to the graticule holder provided in the theodolite box and pulling it out. Grease should be removed by wiping with a chamois leather. Dust can best be removed by means of a strip of thin card slightly narrower than the diameter of the graticule, the strip being torn across leaving a " feathery " edge which can be brushed over the surface. Should the graticule rulings require refilling a piece of soft tissue paper should be blackened with graphite from a soft lead pencil and rubbed over them, the surface being finally cleaned with the strip of card.

(iv) *Illumination system.*—The batteries should be examined for deterioration and replaced at once if they are run down. A theodolite should never be put away with exhausted batteries as serious damage may be done by exuding electrolyte. The battery boxes of the Mark 3 and Mark 4 theodolites must be detached before the instruments are put in their boxes. Lamps should be tested and the adjustment for the graticule illumination should be checked by switching on with the object-glass cap in position and seeing whether the illumination varies when the adjusting knob is turned.

(v) *Tangent screws.*—By means of the tangent screws the telescope should be rotated through 360° in azimuth and 180° in elevation in order to detect, by feeling or listening, any rough places. These can usually be corrected by turning the tangent screw concerned backwards and forwards about the place where the roughness is felt, but if this does not cure the trouble the worm and wheel should be lubricated with a little watch oil which can be introduced when the front cover is taken off (see under backlash adjustments and Plates V and VI).

(vi) *Levelling screws.*—The levelling screws in the base of the instrument should move freely but not loosely. If they are too loose or too tight, they should be adjusted by inserting a tommy bar in a hole in the knurled adjusting ring just above the levelling screw, and tightening or loosening it as required. If the motion of the levelling screws is stiff or rough their threads should be cleaned and greased with vaseline, after first loosening the adjusting rings which should be re-adjusted after this operation.

(vii) *Azimuth clamping screw.*—A check should be made to see that when the azimuth clamping screw is screwed up and the azimuth tangent screw engaged the instrument can only move in azimuth the distance allowed by the spring, relative to the base.

(viii) *Spirit-level.*—The correctness of the mounting of the spirit-level can be checked in the following way. With the azimuth clamping screw N tight and the azimuth tangent screw J released the instrument should be turned until the spirit-level is parallel with two of the base levelling screws, which are then used to bring the bubble to the middle of its tube. The instrument should then be turned through 180°, bringing the spirit-level parallel to its first position but with the ends reversed, and if the bubble is displaced from its central position half the error should be corrected by the capstan-head adjusting screw of the level and half by the two foot-screws. This procedure should be repeated for both positions of the instrument until the bubble remains central in both. Finally the instrument should be turned through 90° from the first position, bringing the spirit-level in line with the third foot-screw, which alone should be used to bring the bubble again to its central position. The bubble should then remain central while the instrument is rotated through 360° about its " observation " axis (see below).

(ix) *Centres or axes.*—There are two vertical axes of rotation. The orientation axis is the one about which the instrument freely turns when the azimuth tangent screw (I, Plate VI) is engaged and the azimuth clamping screw N is released. This is the axis that is used in orientating the theodolite. The observation axis is the axis about which the instrument freely turns when the azimuth clamp N is tightened and the tangent screw I is disengaged. This is the axis about which the theodolite rotates in azimuth when a balloon is being followed.

In a correctly adjusted instrument these two axes should be coincident. The coincidence may be tested by first adjusting the level perpendicular to the observation axis as described in (viii), so that when the instrument is rotated slowly about the observation axis the bubble does not move appreciably from the centre of its travel. Then with the azimuth tangent screw engaged and the azimuth clamping screw released the instrument should be slowly rotated about the orientation axis. If the bubble now shifts, the axes are not coincident and the cones of the bearings probably require regrinding by an instrument maker. No adjustment can be made by the observer to correct for this, but slight lack of coincidence between the axes does not affect appreciably the results obtained with the instrument. It renders relevelling necessary after the theodolite has been set in azimuth before the commencement of an ascent.

(x) *Sights.*—The accuracy of the open sights should be checked by aligning them on a distant object. This will be reasonably near the centre of the field of view of the main telescope unless the sights have been damaged, in which case they will need to be replaced.

(xi) *Coincidence of the two telescope fields.*—To test this the observer should sight a distant object with the main telescope and change over to the auxiliary telescope. The object should appear near the centre of the field of the latter but if it does not the cover on which the small object glass is mounted should be removed. Under the cover is a small capstan-headed screw with a hexagon lock-nut forming an adjustable stop for the mirror. The lock-nut should be loosened and, by means of a tommy bar, the screw turned until the object appears in the centre of either field. The lock-nut should be tightened without moving the screw, and the cover replaced.

(xii) *Backlash.*—This may develop both in azimuth and elevation movement, and is objectionable because it means that two different readings can be obtained at the same position of the telescope. To measure the extent of the backlash a well defined object should be brought on to the centre of the graticule, using the tangent screws, from a distance of 2° or 3° first from one side and then from the other. The difference between the readings is the backlash. It should not exceed 0·03°. In order to avoid serious backlash the tangent-screw mechanisms should be in good adjustment. They will very rarely need attention and no adjustment should be attempted unless it is certain that it is necessary. If it is, the following procedure should be adopted.

The tangent screws are held between pivots (R, Plate VII) in rigid frames which are themselves pivoted to the body of the instrument (Q). They are put into and out of engagement by rotation of the frames about their pivots. Backlash may be caused by end play in either of the pivots or by imperfect engagement of the tangent screws with the worm-wheels. Another possible cause is stiffness in the pivots of the frames or weakness of the springs; this means that the worm is not pressed firmly into the worm-wheel. If there is end play it can be seen and felt by slightly rotating the telescope, or the body of the instrument, with the tangent screw engaged and with the fingers touching the drum. End play between the frame and the body may be taken up by tightening the small steel screw, in the body of the instrument, which forms one pivot of the frame; that for the elevation tangent screw is on the observer's right, that for azimuth on the left. Locking rings are provided which should be slackened before, and tightened up after, the adjustment has been made. These pivots should be only just tight enough to prevent end play.

If this adjustment does not effect a cure, the front cover of the instrument should be removed by unscrewing the four screws fixing it to the body, taking care not to lose the screws or to break the wire to the circle-illuminating lamp. (The wire can be disconnected by loosening the small binding screw behind the lamp socket.) The pivots between the frame and the tangent screw should now be visible and should be adjusted in the manner described above. Before the cover is replaced, the opportunity may be taken to lubricate the tangent screw, worms and all pivots with a very small amount of thin oil. As adjustment of the pivots can alter collimation this should be checked by the procedure described in (xiv) below.

(xiii) *Telescope axes bearings.*—These must be adjusted so that the transit axis of the telescope is horizontal. To test whether this condition is satisfied

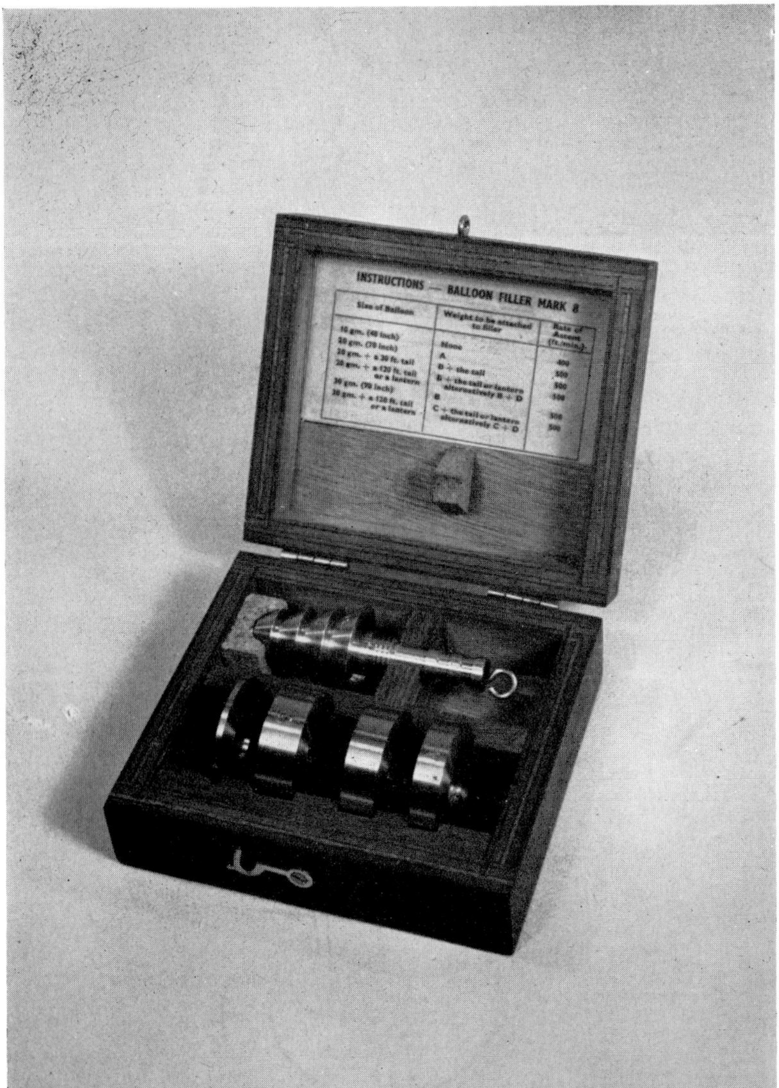

METEOROLOGICAL OFFICE PILOT-BALLOON FILLER, MARK 8

Plate VI

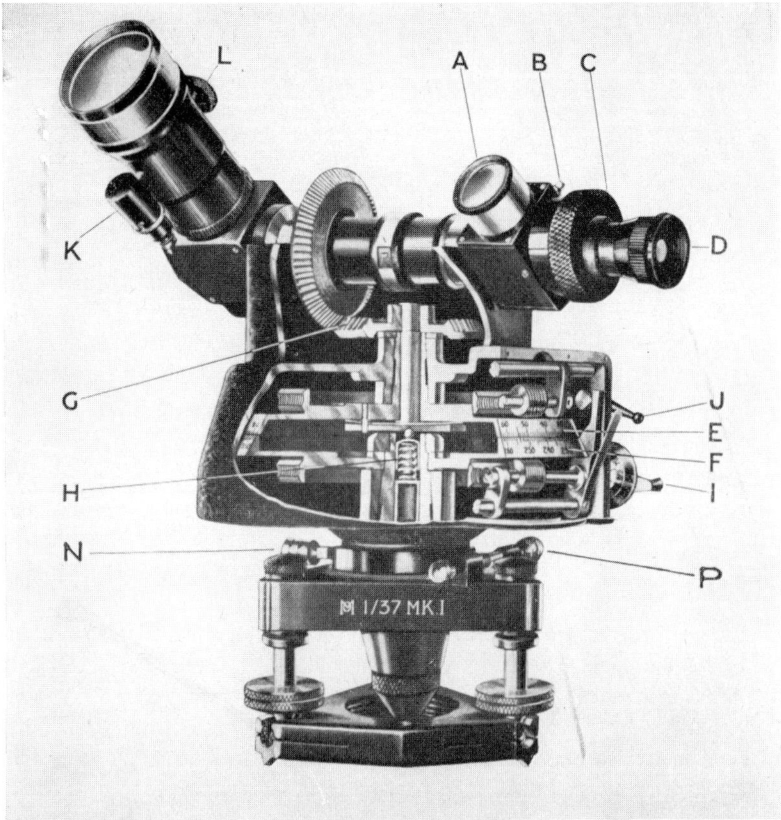

SECTIONAL VIEW OF METEOROLOGICAL OFFICE PILOT-BALLOON THEODOLITE, MARK 1

A, auxiliary sighting telescope
B, lever for auxiliary telescope
C, knurled ring for adjusting focus
D, eye-lens
E, elevation circle
F, azimuth circle
G, bevel gears
H, spring
I, azimuth tangent screw
J, worm disengagement lever
K, holder of bulb illuminating graticule
L, knurled disc for adjusting illumination
N, azimuth clamping screw
P, clamping screw

METEOROLOGICAL OFFICE PILOT-BALLOON THEODOLITE,
MARK 4

TANGENT SCREW MECHANISM OF METEOROLOGICAL OFFICE PILOT-BALLOON THEODOLITE

R, tangent screw pivots
Q, body of instrument

Plate VIII

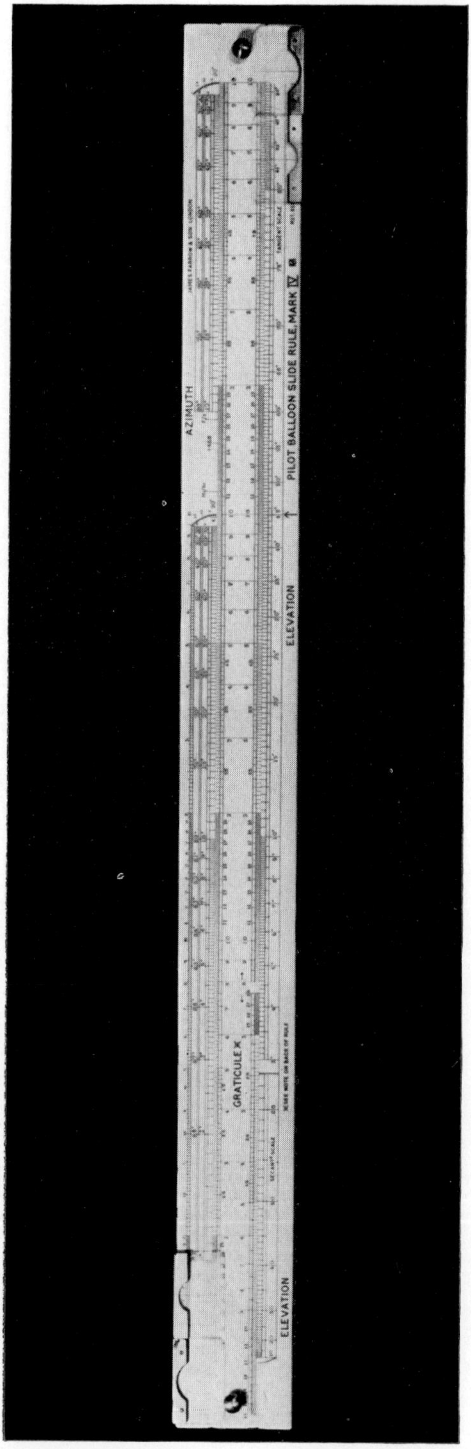

METEOROLOGICAL OFFICE PILOT-BALLOON SLIDE-RULE, MARK 4

the telescope should be directed to a clearly defined point X at an elevation of about 45° and by means of the tangent-screw mechanisms the object should be brought to the centre of the graticule. The telescope should then be depressed as far as possible and a point *Y*, provided, for example, by a scale or levelling staff laid on the ground at some distance, noted to be coinciding in the field of view with the centre of the graticule. The telescope should be turned through 180° in azimuth and redirected at the lower point *Y*. If, after elevating the telescope the graticule centre again coincides with the first point X the transit axis is horizontal. If there is lack of coincidence the instrument is out of adjustment. The error may be disregarded if it is less than 1·5 divisions on the graticule. If it exceeds this the theodolite will require the attention of an instrument-maker, since grinding of the bearings may be necessary.

(xiv) *Horizontal and vertical collimation.*—The line of collimation of the telescope should coincide with the transit axis of the telescope. It may be displaced in the horizontal plane and in the vertical plane. To test for any displacement a distant object of low elevation should be sighted and the azimuth and elevation readings taken. The tangent screws should then be unclamped and the telescope turned through 180° in azimuth and again sighted on the same object. After allowing for the 180° rotation the azimuth and elevation readings should agree with the previous readings to within 0·2°. Errors exceeding this amount are most probably due to a shift of the pentagonal prism and would require correction by an instrument-maker. It should be noted that in one of the two positions the elevation drum reading must be subtracted from 1·0 to give the correct reading, e.g., if the two elevation readings appear to be 2·56° and 2·38°, the actual readings are either 2·56° and 2·62° or 2·44° and 2·38°, with a true difference of 0·06°. Further, if the object sighted is less than half a mile away, allowance must be made for the fact that the main object glass is offset from the vertical axis; this involves a small addition to one azimuth reading, and an equal subtraction from the other. For an object a quarter of a mile away it amounts to about 0·02 degrees.

As part of the collimating adjustment it may be necessary to move the glass windows of the scales sideways, after loosening the screws in their frames, to bring the reference line on to a whole-degree graduation when the drums read zero. To obtain correct adjustment for azimuth and elevation simultaneously may require a slight rotation of the drums with respect to the worm-spindles, which can be done when the screw at the centre of the drum is loosened.

3.4. TAIL AND DOUBLE-THEODOLITE METHODS AND NIGHT ASCENTS

3.4.1. Tail method

The disadvantage of the single-theodolite method of pilot-balloon observation is the necessity for assuming a rate of ascent of the balloon. It is overcome to a certain extent by the tail method. This consists in observing the apparent length of a tail attached to the balloon, as seen through the telescope of the theodolite. The tail method is useful when the wind is strong enough to keep the angle of elevation of the balloon below about 40°. At higher elevations the method is not applicable.

The trigonometry of the method is shown by Fig. 1 in which E is the angle of elevation and h the height of the balloon, l is the length of the tail attached to it and θ the angle subtended by the tail at the observer's eye. The height is given by:

$$h = \frac{l \cos E \sin E}{\theta} = \frac{l \sin 2E}{2\theta} .$$

Thus, measurement of the angle subtended by the tail enables the height of the balloon to be calculated without making any assumptions about the rate of ascent of the balloon. The method assumes that the tail remains vertical whereas it is usually swinging. A single observation is therefore subject to large errors and it is necessary to smooth out the irregularities of consecutive readings.

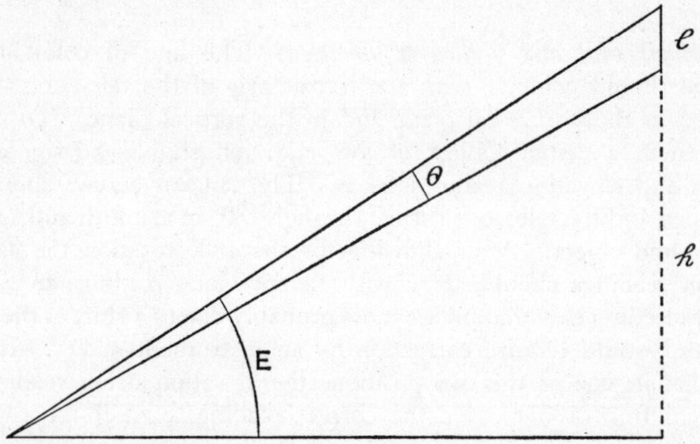

FIG. 1—TRIGONOMETRY OF TAIL METHOD OF PILOT-BALLOON OBSERVATION

The graticule.—The angle θ is measured with the aid of a graticule placed at the focus of the theodolite telescope. The graticule consists of a glass scale with divisions formerly every 0·01 inch but now corresponding to 1/1000 radian, against which the apparent length of the tail can be estimated. If the apparent length is m graticule divisions and k divisions are equal to an angle of 1 radian, then $\theta = m/k$ and

$$h = \frac{kl \sin 2E}{2m} .$$

The graticule scale value k is a constant of the theodolite and can be determined by setting up a foot-rule vertically at a measured horizontal distance of 100 ft. from the object glass of the theodolite. The centre of the rule should be in the same horizontal plane as the eyepiece of the telescope. The rule is brought into accurate focus and its length determined in terms of scale divisions (n) of the graticule. Since the angle subtended by the rule is 0·01 radian the value of k will be $100n$.

Form of tail.—The tail consists of a length of thread the end of which is marked by a pendant formed by a foolscap sheet of paper having a length of thin aluminium wire gummed along one of the shorter edges, while the opposite edge of the paper is folded over and gummed to assist in keeping the paper fairly rigid. The end of the thread is tied round the aluminium wire through a small hole in the paper. The visibility of the pendant is often improved by the use of coloured paper selected according to the colour of the sky. Silvered paper is rather better than white paper with a clear sky, red is suitable for a hazy or cloudy sky, and black for an overcast sky. A form of tail which spins is to be avoided, because it causes the thread to shorten and gives values of heights which are too high. The addition of the tail will make a difference to the free lift required to produce the standard rate of ascent (see Section 3.1.3).

Length of tail.—In order to simplify the computations it is necessary to make the length of the balloon tail bear a definite relation to the value of the graticule factor k. This relationship is given by the formula

$$l = 1{\cdot}2 \times 10^5/k\,,$$

where l is the length of the tail in feet. The standard length thus calculated will be about 100 ft. and once it has been determined two nails can be fixed in a wall at an appropriate distance, say $l/20$ apart and the tails can be measured off by winding 20 such lengths on the nails. In the early stages of an ascent all the tail may not be visible in the field of view of the telescope. This difficulty can be overcome by attaching a subsidiary pendant at a quarter of the full tail length below the balloon, and multiplying the apparent length to this pendant by four.

Operational procedure.—The launching of balloons with a tail is facilitated by the use of a release device. This generally consists of a stout cardboard drum with wooden ends mounted on a spindle about which it can rotate freely. The pendant is first wrapped round the drum and then the thread, care being taken that the ends of the aluminium wire are free from the thread. The inflated balloon is tied to the free end of the thread and the drum is now held so as to prevent it rotating until the balloon is being released. The balloon is then liberated by allowing the drum to rotate.

3.4.2. Double-theodolite observations

To avoid the necessity for making any assumptions about the rate of ascent of the balloon the latter may be tracked by two theodolites sited at the ends of a base line. The trigonometry of the method is straightforward, being the solution of triangles having a common side representing the height of the balloon, which can therefore be determined precisely from the angular measurements and the length of the base line. It is necessary to adopt certain precautions if the full advantage of the method is to be gained. A base line of length not less than about one fifth of the maximum range to be observed is desirable and it should be in a direction roughly at right-angles to that of the prevailing wind. It should be very accurately surveyed and if it can be selected so that the two ends from which the observations are to be taken are at the same elevation the computations will be simplified. The orientation of the two theodolites must be made very accurately; this can be conveniently done by setting the telescopes along the known direction of the base line, with the azimuth circles reading 360°. Another important point is the need for exact synchronization of readings; this is best achieved by having telephonic communication between the two stations.

3.4.3. Night ascents

Paper lanterns.—When making pilot-balloon ascents at night it is necessary to show a small light below the balloon. The most economical source of illumination is a paper lantern containing a small candle. The method of construction of the Meteorological Office type of lantern (Stores Ref., Met. 140) is described in Section 10.2.6 of Part I. It consists of a collapsible tissue-paper cylinder about

5 in. high, with thin cardboard discs about 4 in. in diameter as ends, the upper disc having a circular hole about 2 in. in diameter and a loop of soft thin wire attached. A length of about 2 in. of " Christmas " candle is stuck on to the centre of the lower disc inside the lantern and provides the necessary illumination. The lantern is suspended from the balloon by about 5 ft. of thread but since a sudden jerk during launching is liable to extinguish the light it is advisable, and in strong winds essential, to include in the suspension about 2½ ft. of bare elastic (cross-section about 0·03 in.). It may be necessary to put a cap over the opening of the lantern if a balloon ascending at 1000 ft./min. is used.

In practice the maximum heights observed with paper lanterns on clear nights are of the order of 20,000 ft. Some improvement on this has been obtained by using short lengths (1 in.) of household candles instead of the thinner " Christmas " candles.

Electric torches.—No great advantage is obtained by using electric lamps supplied from dry batteries, since the latter have a poor shelf life and their efficiency falls off at low temperature. Electric lamps powered from water-activated batteries, however, are much more suitable for high ascents at night. An electric torch for pilot balloons (Stores Ref., Met. 1674) recently adopted by the Meteorological Office consists of two lead dioxide–magnesium cells with a 2·5 volt, 0·3 amp. bulb attached. The battery remains inert until soaked in water and has a long shelf life when kept dry. A plastic bag (Stores Ref., Met. 1780) is provided for protecting the battery from much loss of heat in the air flow during an ascent. The method of using the torch is first to fill a balloon to have an extra free lift of 75 gm., which is twice the weight of the torch, and tie a short length of string to the neck. Then a hole is cut in the bottom of a plastic bag just large enough for the lamp bulb to pass through and the bag is fitted over the battery with the bulb through the hole and the supporting loop passing through the mouth of the bag. The battery is soaked in water for half a minute and the excess water drained off. The mouth of the bag is then tied up and the torch attached to the balloon ready for launching immediately. It has been found possible to obtain ascents with 100 gm. balloons up to 40,000 ft. on many occasions with this type of torch which produces a light of the order of 1 c.p. for at least 30 min., whereas the light from a " Christmas " candle is about 0·25 c.p.

3.5. COMPUTING DEVICES

In any of the three methods of pilot-balloon observation (single-theodolite, double-theodolite and tail method) the computations required are simple but they are numerous as they involve the solution of many triangles. Several methods of dealing with the computations, many of them graphical, have been devised. There are, however, advantages in using a slide-rule for such a purpose; an ordinary slide-rule will serve if the sines and tangents are referred to the same base. (In most rules the scales with which the sines and tangents are compared are not the same.) There are, however, features in pilot-balloon computations which make the use of a special rule desirable.

3.5.1. Pilot-balloon slide-rules

Since the angular co-ordinates of the balloon change but little from one observation to the next, while the linear co-ordinates change considerably, it is advantageous to have the trigonometrical scales of the rule fixed and the linear scales movable.

Two sine scales are required as the sine and cosine of the same angle must be indicated by cursors simultaneously. There must be three cursors at least to work on the trigonometrical scales and they must be made so that those on opposite sides of the rule may pass each other readily. These features were all introduced in the original pilot-balloon slide-rules made for the Meteorological Office in 1915. These rules were provided with double slides with which allowance could be made for variations in the assumed rate of ascent of the balloons. With the adoption of standard rates of ascent the extra slide became unnecessary and observers were instructed to lock the two slides together. When the use of the tail method became more general the design of the slide-rule was modified to incorporate an additional pair of scales and a fourth cursor for the calculations involved in tail ascents. The modified rule (Mark 2) was described in detail by F. J. W. Whipple[42].

Meteorological Office pilot-balloon slide-rule, Mark 4 (Stores Ref., Met. 868).—This slide-rule is the model in current use at Meteorological Office stations. Its main difference from the Mark 2 model is that the stock and slider, which are 24·5 in. long, are made of ivory perspex, or similar plastic material, instead of wood. From the illustration in Plate VIII it will be seen that there is a logarithmic scale, three times repeated and thus covering a range from one to 10^3, on both the upper and lower edges of the slide. The upper part of the stock carries two pairs of sine and cosine scales, one pair extending from 0·5° to 90° and the other from 10° to 90°; the degrees from 0° to 20° on each cosine scale are marked on a small arc instead of being crowded together as in an ordinary slide-rule. A logarithmic scale, twice repeated, and identical with part of the slider scales, is also provided on the upper part of the stock; this permits of the rule being used in a limited way as an ordinary slide-rule. The lower part of the stock has a tangent scale on the right extending from 3° to 84·3° and a scale of squares of secants from 0° to 63° on the left, the interval between 40° and 50° on this latter scale being twice that between 40° and 50° on either of the upper sine and cosine scales. A graticule scale, of reciprocal logarithms, replaces part of the logarithmic scale on the lower left part of the slider. Fiducial marks " K ", " Mi/hr ", " 1·468 " and " f/s " are engraved on the upper part of the stock, between the main trigonometrical scale and the subsidiary one, for the conversion of units as explained below.

The general relations used in operating the slide-rule to obtain the height of the balloon h, in feet, the total horizontal distance travelled D, in hundreds of feet, and the easterly and northerly components D_E and D_N of this distance from the measurements of azimuth A, elevation E and apparent length of the tail m in graticule units are:

$$h = \frac{kl}{m} \sin 2E = D \tan E$$

$$D = \frac{kl}{m} \frac{1}{\sec^2 E}$$

$$D_E = D \sin A \quad \text{and } D_N = D \cos A\,,$$

where k is the number of graticule divisions per radian, l the length of the tail in feet (see Section 3.4.1), and m its apparent length in graticule units. Normally the product kl is arranged to be $1 \cdot 2 \times 10^5$. If a different value is used the graticule readings must be multiplied by $1 \cdot 2/kl$ before calculation on the rule.

In using the slide-rule the normal procedure is first to set the left-hand lower cursor at the observed value of the elevation E on the secant² scale and to move the slide so that the value of m on the graticule scale is against $\sec^2 E$. Then, after setting the right-hand lower cursor at the observed elevation on the tangent scale the height is read from the logarithmic scale on the slide. The upper cursors are now set at the observed azimuth on the sine and cosine scales and from these cursors the values of D_N and D_E are read on the logarithmic scale on the slider. If the value of D is required it can be read from the logarithmic scale opposite sin 90°. Values of the wind components are obtained from the differences of successive values of D_N and D_E. Then the larger of the components (irrespective of sign) is set at tan 45° on the tangent scale and the reading from this scale against the smaller component gives the angle (less than 45°) between the path of the balloon, in the minute interval concerned, and the nearest of the cardinal lines (W.–E. or S.–N.) bounding the quadrant in question. Finally, with this angle on the sine scale set against the smaller component on the logarithmic scale, the value opposite sin 90° gives the horizontal distance travelled by the balloon in the minute interval, and thus the wind speed in units of 100 ft./min. Without further adjustment, the speed may be read off in knots (from the fiducial mark at " K "), in miles per hour (at " Mi/hr ") or in feet per second (at " f/s "). Inter-unit conversions may also be made by direct reading from these fiducial marks. The mark " 1·468 " is for conversion from miles per hour to feet per second.

If the observations are made on a moving ship account must be taken of the ship's course and speed. This is most conveniently done by adding algebraically the components of the ship's velocity to the components of the relative wind velocity.

The slide-rule should always be used with care and not handled with soiled or stained fingers. If it is necessary to clean the surface warm soapy water should be used. Abrasives, petrol or solvents must never be used.

3.5.2. Graphical methods

Pilot-balloon computations may be carried out entirely by plotting on a polar co-ordinates chart, with one quadrant ruled also with rectangular co-ordinates. The latter enable the plan ranges, $h \cot E$, of the balloon to be obtained from the elevation and height, after which the plan positions of the balloon can be plotted at these distances along the appropriate azimuths on the polar chart. Wind speed and direction are then obtained, with a scale and protractor, from lines joining successive points. In the case of a two-theodolite ascent the plotting may be done on a pair of polar diagrams whose centres are separated by a distance corresponding to the theodolite base line on the scale selected for the graph. Various mechanical devices have been designed to facilitate graphical methods. A good example is one devised by L. W. Pollak[43]; it has the advantage of being suitable for tail ascents as well as for ascents of assumed constant rate.

Pollak's pilot-balloon plotting instrument.—This instrument, which is illustrated in Fig. 2, consists of three parts and it is used on a plotting chart which has a circular scale of 360°. One part is a quadrant, OCB, pivoting at the centre of the plotting chart and marked with a scale of degrees on the outer edge of its arc and with a scale of double angles from 0° to 45° on the inner edge. The radial arm OC carries a scale for m, the apparent length of the balloon tail in graticule units, and is extended beyond the arc. The second part of the instrument, AOE, is in the form of two arms fixed at right-angles and pivoting at their intersection at the centre of the chart (above the quadrant). The longer arm OA serves for setting the azimuth on the polar chart while the shorter arm OE is used for setting the elevation on the quadrant. The third part, MPH, is also in the form of a right-angle, the narrower arm, PH, of which has a scale of heights while the wider arm, PM, has a scale of *m* on its inside edge and a scale for wind speed on its outside edge. This part slides along the longer arm of the other right-angled member.

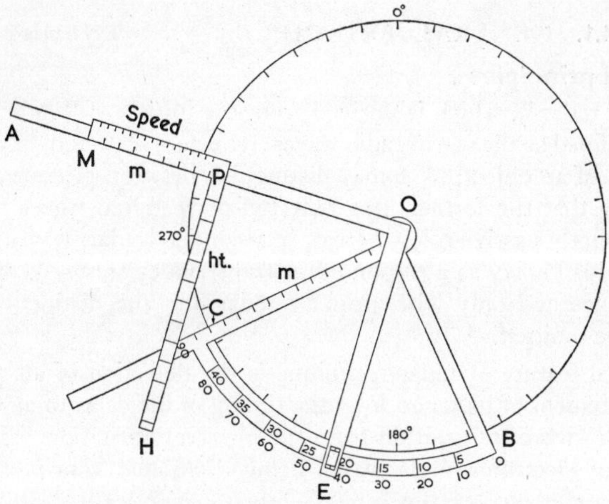

FIG. 2—POLLAK'S DEVICE FOR EVALUATING PILOT-BALLOON ASCENTS

The method of use is first to set the azimuth arm OA to read the observed azimuth on the circular scale of the plotting chart. Then, keeping that arm fixed, the quadrant is rotated until its outer scale shows the observed elevation at the end of the shorter arm OE. The sliding piece is moved along the azimuth arm until the height scale on PH intersects with the radial edge of the quadrant at the assumed height of the balloon. The intersection of the height scale with the azimuth arm at P then indicates the plan position of the balloon. This is plotted on the chart and from successive plan positions the wind speed is obtained with the aid of the scale on the outer edge of PM and the direction with the aid of set-squares or protractor.

In the case of a tail ascent, after setting the azimuth arm the quadrant is rotated until its inner scale shows the observed elevation at E. The sliding piece is then moved until the outer edge of the height scale PH intersects the scale on the quadrant OC at the observed value of *m*. The plan position is then indicated at the observed value of *m* along the scale of the slider PM, while the balloon height is read directly from the height scale PH.

The theory of Pollak's device can be readily derived from Fig. 2 and the only point that need be mentioned here is that the *m* scales are based on the quantity $kl/2m$ in the formula for the horizontal displacement D of the balloon in terms of the double angle of elevation (shown on the inner edge of the quadrant arc), thus:

$$D = \frac{kl}{m}\cos^2 E = \frac{kl}{m}(1 + \cos 2E).$$

CHAPTER 4

PRIMARY RADAR FOR UPPER WIND MEASUREMENT

4.1. GENERAL PRINCIPLES AND THEORY

4.1.1. General principles

The term " radar ", which is derived from the initials of " radio detection and ranging ", is defined as the use of radio waves, reflected, scattered, or re-transmitted, for the location of an object. We may distinguish between primary and secondary radar by noting that the former uses reflected or scattered waves from an object which plays a purely passive role whereas, in secondary radar, waves re-transmitted from an apparatus known as a responder or transponder are used. In this chapter we shall be concerned only with primary radar and the distinction " primary " will generally be omitted.

The essential feature of radar technique is the detection of an object and the accurate measurement of its range R by the timing of the delay of an echo, from the object, of a radio wave transmitted from the observation station. The velocity of propagation c of electromagnetic waves being accurately known, the delay time $2R/c$ provides a measure of the range. Since c is $2 \cdot 9979 \times 10^{10}$ cm./sec. the measurement of ranges up to 100 km. with an accuracy of $\pm$ 20 m. involves the measurement of time intervals up to 670 μ sec. with an accuracy of $\pm$ $0 \cdot 13$ μ sec., i.e. to one part in 5,000. Such precision is quite readily obtainable with radar and has, in fact, been improved upon considerably.

One of the methods of timing the echo delay is by using frequency modulation of a carrier wave. In this system the echo signal arrives back with a radio-frequency differing from that of the wave being transmitted at the same time and the resulting beat frequency from the superimposed oscillations is a measure of the echo delay. The method is not suitable for use with fast-moving targets since the change of frequency due to the Doppler effect may be comparable with that produced by the modulation. The Doppler principle does, in fact, itself provide a means of measuring the radial velocity of a fast-moving target. Frequency modulation is used in radar altimeters for aircraft, for which the sea or land surface acts as a relatively static target, but for other purposes the principle of pulse transmission is generally used. In this, discrete bursts or pulses of radiation are transmitted and the time interval separating individual pulses and their echoes is observed. For the pulse techniques, therefore, the receiving apparatus must incorporate a suitable time-measuring device, preferably calibrated to give a direct reading of range.

It should be noted that range cannot be measured if the echo pulse begins before the transmitter pulse is finished. Thus if the duration of the pulse, generally known as the pulse width, is t, the minimum range is not less than $ct/2$, which corresponds to 150m. for a pulse width of 1 μ sec. At least one edge of the pulse, preferably the leading edge, must be sufficiently steep for the time taken in forming it to be less than the permissible error. The pulse method necessitates some means of displaying the transmitted and echoed pulses, together with a reference standard of time (or frequency). Moreover, the display system must be synchronized with the transmitter.

For a radar equipment to be suitable for accurately locating an airborne object it must not only be capable of measuring the range but must also provide means of measuring the azimuth and angle of elevation. This necessitates an aerial system having directional properties. The radiation is concentrated into a beam and, as in the case of searchlights, improvement in angular accuracy can be obtained by decreasing the width of the beam. There is, however, a limit to this, since too narrow a beam may make the detection of an object very difficult. Some method of beam-switching, to be described later, is therefore used in precision radar and some equipments are provided with a means of following the target automatically.

There are various factors that govern the choice of the radio-frequency to be used in a radar equipment but it will suffice here to mention two that set limits to the range of choice. If frequencies lower than 600 Mc./s. are used reflections from the ground may cause large errors. The other extreme is limited by the fact that at frequencies greater than about 5000 Mc./s. serious reduction of detection range can occur in heavy rain (J. A. Saxton[44]). An advantage of choosing frequencies near the upper limit is that they allow the production of narrow beams without the use of large aerials.

4.1.2. Basic theory

The two fundamental aims of radar technique are, as already stated, the detection of a distant object and the determination of its position. For the measurement of range the time taken for a pulse of radio energy to travel out to the object and back again is observed, and then expressed as a distance by using the constant velocity of electromagnetic waves in free space (now taken as 299,792 km./s.). The fact that the velocity in the atmosphere is slightly lower is usually ignored, since the difference, which varies with the radio refractive index of the air and therefore with humidity, temperature and pressure, averages about 35 parts in 100,000 at the earth's surface and about one part in 10,000 at an altitude of 10 km. Smaller variations about the surface average of up to approximately plus or minus one part in 10,000 are found although at any one place the total seasonal variation is about one half and the daily variation up to about one quarter of this range. For high-precision measurements at long range and low elevation angles the effect of refraction can become significant. For example, an accuracy of $\pm$ 10 m. at a range of 200 km., which is practicable with modern radar technique, corresponds to an accuracy of five parts in 100,000. There would be no advantage in using this high precision unless the effects of atmospheric refraction and its variations were taken into account. The variation of radio refractive index with altitude causes a ray to follow a curved path, convex upwards. While the effect of this curvature is quite negligible in range measurement its effect upon the apparent elevation angle can be significant. This has been considered in detail by Hooper and Taylor[45] who present for a standard atmosphere a graph of errors within the limits normally experienced in wind-finding. Table VI illustrates the general magnitude of the effect. The effect of extreme departures from the standard atmosphere assumed is reported to be of second order or less in the European area and rather greater in the subtropics.

TABLE VI—CORRECTION TO APPARENT ALTITUDE, DUE TO REFRACTION

Apparent altitude	Elevation angle 7°	10°	14°	20°
km.		*metres*		
10	− 95	− 45	− 30	− 15
20	− 265	− 140	− 75	− 40
30	− 435	− 235	− 130	− 65

In considering the problem of obtaining detectable echoes from a distant object we may start with the power P_t (usually in terms of the peak value in a pulse) which a transmitter delivers to its radiating aerial. If the latter were isotropic the intensity of the radiation at a range R would be $P_t/4\pi R^2$. In practice the transmitting aerial is arranged to be directional and it will have a gain G_t in the most favoured direction which may be defined as the ratio of the maximum intensity produced at a given distance by a given total radiated power to that produced at the same distance by the same power radiated isotropically. The effect then, is to cause an intensity I_t in the most favoured direction given by:

$$I_t = \frac{P_t G_t}{4\pi R^2}.$$

Of the power incident on a target only a fraction will be scattered or reflected back along the direction of incidence. The echoing efficiency is specified in terms of the "echoing cross-section", σ, of the target for a particular direction (and polarization) of incident and scattered or reflected waves. The intensity I_r of the radiation echoed back to the receiver is therefore given by:

$$I_r = \frac{\sigma I_t}{4\pi R^2}.$$

If the effective aperture of the receiver aerial is A_e, the power P_r absorbed at the receiver input is $A_e I_r$, or

$$P_r = \frac{\sigma A_e I_t}{4\pi R^2} = \frac{\sigma A_e P_t G_t}{16\pi^2 R^4}.$$

It can be shown that the gain of an aerial is related to its effective aperture, thus

$$G = \frac{4\pi A_e}{\lambda^2},$$

where λ is the wavelength of the radiation and, therefore, if we express A_e in terms of the gain G_r of the receiver aerial the echo power received may be written:

$$P_r = \frac{\sigma G_r G_t P_t \lambda^2}{64\pi^3 R^4}.$$

This shows that if P_r is the minimum energy that can be detected by the receiver the maximum range obtainable is proportional to the fourth root of the transmitter power. If the receiver and transmitter aerials are of the same dimensions and design then $G_r = G_t$ and the maximum range is proportional to the square root of the gain of the aerial. For aerials of similar type, therefore, the maximum range is proportional to their linear dimensions.

In considering the variation of maximum range with wavelength, performance changes in both the ground radar and the airborne reflector must be taken into account. Since the system is inevitably marginal in performance at extreme range both the radar aerials and the reflector will be made as large as is reasonable from structural and economic considerations. For the type of reflectors (other than a sphere) normally used in upper wind measurement the echoing cross-section is inversely proportional to the square of the wavelength (Section 4.7.3) so that when, as before, the transmitting and receiving aerials are of similar type, we have:

$$R^2 \propto \frac{A_e}{\lambda^2}\sqrt{\frac{P_t}{P_r}}.$$

Thus for a given size of reflector and ground aerials the maximum range is inversely proportional to the wavelength.

The minimum power detectable by the receiver is usually limited not by the amplification that is available but by " noise " arising from thermal agitation in conductors and by the irregular nature of electronic emission in thermionic valves. Generally a weak echo can only be detected if its power is of the same order as that due to noise, though by pulse integration technique it is now possible to detect signals in which the individual pulses have only a small fraction of the noise power. Since noise is distributed over an infinite frequency spectrum the noise power in the output of the receiver is directly proportional to the range of frequency accepted and amplified (the receiver band width). Requirements of range accuracy, however, may make it impracticable to improve the ratio of the signal to noise powers, and therefore the maximum range, by decreasing the band width. Range discrimination determines the pulse width but for high accuracy of range measurement it is the steepness of the leading edge of the pulse that is more important. Each pulse consists of a train of oscillations at the same radio-frequency and an ideal pulse would be one with a rectangular envelope enclosing individual oscillations all of the same amplitude, the width of the rectangle being the pulse width. In practice, however, the pulse has a frequency spectrum spread about a mean value and, owing to the time taken for the oscillations to grow and decay, the shape of the pulse envelope is more like that of a cocked hat. The width then has to be defined arbitrarily as, for example, the period during which the pulse envelope exceeds $1/\sqrt{2}$ of its maximum value. The receiver must be capable of passing the main part of the frequency spectrum and for high range accuracy it is important that the band width of the receiver should be adequate to pass the echo pulse without distorting its shape, especially at the leading edge. It should, in fact, equal or exceed the reciprocal of the pulse width. Thus, for a pulse width of 1 μ sec. a band width of about 1·5 Mc./s. would be appropriate if high accuracy of range measurement were required.

The number of pulses per second is known as the pulse recurrence frequency. It is limited at one extreme by the need to allow sufficient time for a pulse to travel to the maximum range and back and at the other by the need for averaging the received signals over short periods of time. The pulse recurrence frequency is generally of the order of several hundred per second.

4.1.3. Microwave pulse-radar equipment

The principal units of a pulse-radar set are shown as a block diagram in Fig. 3. The transmitter and receiver form the main subdivision but each may share some units, such as power supply and aerial. The function of the transmitter is to emit short powerful pulses of radiation in rapid succession and to direct the energy into a restricted beam. It comprises a modulator, a radio-frequency oscillator and an aerial system. The modulator is a device for taking power from a source of supply, storing it and releasing it at the required recurrence rate in the form of voltage pulses of a suitable amplitude, duration and shape. In order to generate the pulses it is necessary to perform a switching operation and the switch must be able to pass high currents and withstand high voltages. Hydrogen-filled thyratrons are frequently used for this operation but have the disadvantage of taking several minutes to warm up. A spark-gap type of modulator is sometimes used; it comes into operation immediately. The pulses may have an amplitude of the order of 10 kV. and a width of the order of 1 μ sec. They are applied at a rate of several hundred per second to the cathode of a special type of ultra-high-frequency valve oscillator, known as a magnetron, in which the current is controlled by a magnetic

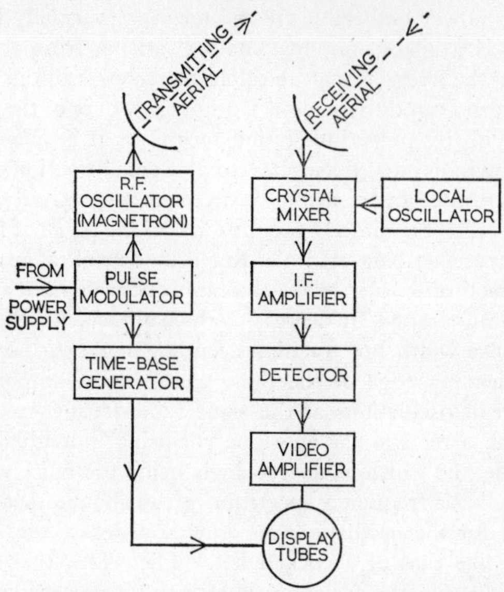

FIG. 3—MAIN UNITS OF A PULSE-RADAR SET

field. The main features are illustrated in Fig. 4. The anode A is a short thick-walled cylinder with a number of built-in resonant cavities and it surrounds the cathode C which is in the form of a tube. The whole is situated between the poles of a magnet whose field causes the electrons to describe curved paths and as they pass the cavities they are velocity-modulated and subsequently bunch together. In striking the anode in bunches they build up oscillations in the resonant cavities. The magnetron valve oscillates at a frequency of thousands of megacycles per second (corresponding to wavelengths of the order of centimetres) and delivers pulses of these oscillations with a peak power of hundreds of kilowatts. The pulses are fed through coaxial conductors, or waveguides, to the transmitting aerial. Provision is also made for the modulator to produce, at the same time as the transmitted pulse, a locking synchronizing pulse which is passed to the receiver where it is used to trigger off the time-base for measuring the echo delay time.

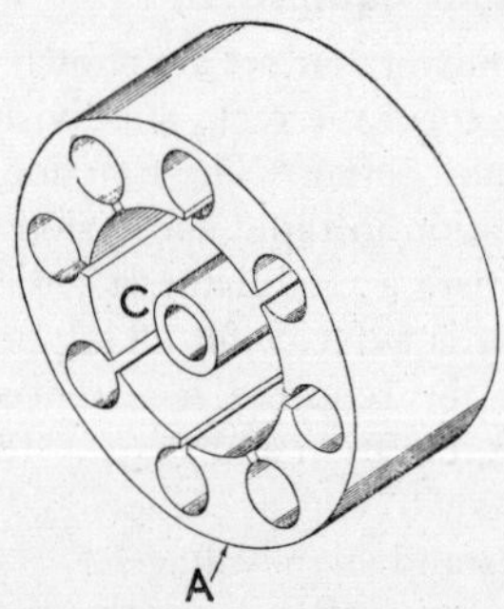

FIG. 4—SECTION OF A CAVITY MAGNETRON

For uses such as wind-finding in which the accurate tracking of airborne objects is involved, a highly directional type of aerial is required and usually takes the form of a dipole at the focus of a metal paraboloid (mirror or " dish "), the dipole being a rod whose length is a simple fraction (usually half) of the wavelength.

The directional characteristics of such an aerial, as shown by a polar diagram of the power distribution around it, take the form of a narrow beam with a maximum of power along the axis of the paraboloid. The beam width may be defined as the angle of the cone at which the power at a given distance bears a specified ratio, usually one-half, to the power at the same distance on the axis. For a paraboloid aerial system the beam width is a few degrees. There must be a means of steering an aerial used for tracking and this may be done manually or automatically. In the latter case the aerial is turned by electric motors which are controlled by the signals received back from the target. The aerial system of the transmitter may be shared with that of the receiver or it may be a separate radiator. With a common aerial system special devices, known as T.R. boxes, using gas-filled valve switches, are needed to ensure that the receiver is protected from the outgoing high-powered pulses.

In the receiving system of the radar set the main units are the aerial, the radio receiver, a time measuring device and the display or presentation unit. The power of the echo signal is relatively small and must be amplified considerably before it can be used to give suitable visual indications. At frequencies of 3000 Mc./s. and above there are difficulties in obtaining satisfactory amplification at the radio-frequency and it is generally the practice to convert the radio-frequency signal into one on a much lower frequency, on which amplification can be performed more efficiently. The type of radio receiver generally used is the superheterodyne, which involves generating at low power a frequency close to that of the received signal and beating one against the other, thus producing an output at a frequency between about 10 and 100 Mc./s. This conversion from an ultra-high to a lower frequency (usually referred to as the intermediate frequency) is effected by a crystal mixer, to which local oscillations are supplied by a valve oscillator of special type. The output of the mixer is amplified by an intermediate frequency amplifier, detected by a normal diode valve, and then, after further amplification by a video amplifier, fed to cathode ray tubes in the presentation unit.

For the measurement of range the presentation unit must include some means of displaying the transmitted pulse and its echo against a frame of reference in time that is synchronized with the operation of the transmitter. This is done by using a time-base generator which is a device, generally incorporating a cathode ray tube, for producing a voltage or a current varying with time. It is triggered by the synchronizing pulse from the modulator of the transmitter. A crystal-controlled valve oscillator provides a suitable means of calibrating the time-base. If r is chosen as a suitable scale interval of range the equivalent frequency is given by $c/2r$, where c is the velocity of propagation. For a range interval of 1 km. this gives an equivalent frequency of 149·896 kc./s. and one cycle of oscillation corresponds to 1 km. displacement on the cathode ray tube. It is not practicable to have a trace on the tube that is long enough to cover a large range, with a sufficiently open scale for accurate measurements. For this purpose a specially rapid time-base, on a separate tube, may be used to display a selected small section of the range scale. This is triggered by a specially generated delayed pulse and the selected part of the trace may be brought into line with a reference mark on the screen of the tube by controlling the delayed pulse with a calibrated potentiometer, the setting of which will be a measure of the range. The adjustable delayed pulse may also be used as a means of correlating any signal on the range scale with indications on the direction tubes. This is done by applying the delayed pulse as an extra brightening voltage to the control electrodes of all the cathode ray tubes. The resulting bright patch, known as a " strobe " or " gate ", on the range tube is adjusted to overlap the selected echo, which is therefore brightened on the azimuth and elevation tubes.

For equipment designed for accurate tracking of airborne objects the receiving aerial is either shared with, or is similar to, the transmitting aerial. In the latter case, the two aerials are linked mechanically so as to point always in a common direction. As an indication of the accurate setting of an aerial system on to a target the criterion of maximum response can be used but it requires a beam width of less than 4° and the beam must be traversed repeatedly across the target to determine the angle of maximum response. Wider beams, although resulting in a smaller amount of energy falling on the target, can be used with a beam-switching, sometimes known as " split-beam ", technique. This is illustrated in Fig. 5. In this method the axis of the beam is arranged to oscillate through a small angle. From the polar diagram of the beam at opposite extremes of the oscillation shown in Fig. 5 it will be seen that the strength of the response from an object A is proportional to OX when the axis of the beam is along OP and to OY when it is along OQ. When the object is on the line bisecting the angle between the two positions of the beam the echoes are of equal strength and show as " pips " of equal amplitude on a cathode ray tube. It should be noted that for a common transmitting and receiving aerial the principle applies twice over, so that the effective beam width of such a system is smaller than for separate aerials with only the receiving aerial oscillating. If both azimuth and elevation are to be measured the beam is arranged to perform a conical oscillation (by rotating the dipole receiver aerial in its reflector) and by suitable timing and switching arrangements only those signals received when the beam is deflected to the right or left are displayed on the azimuth tube, while only those received when the beam is tilted up or down are displayed on the elevation tube. By means of handwheel or automatic control and motor driving gear the aerial can be turned until the pairs of echoes on each tube are equal; with the split-beam method, therefore, at least four pulses must be radiated during one aerial switching cycle.

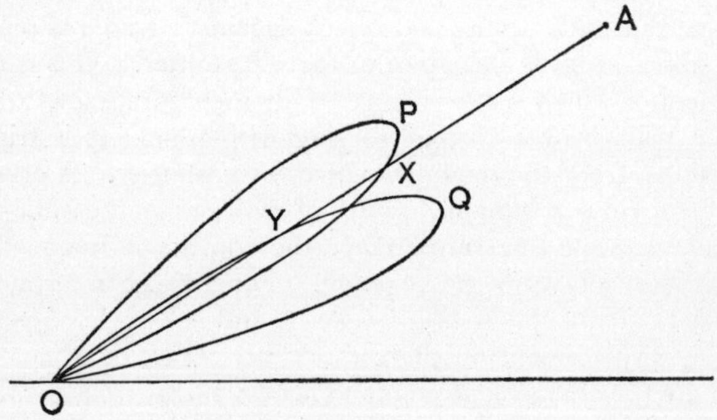

FIG. 5—BEAM-SWITCHING (SPLIT) METHOD FOR ACCURATE DIRECTION-FINDING

4.1.4. Application to wind-finding

Principal design requirements.—The first consideration in the design or adaptation of a radar set for measuring upper winds is that it should have adequate performance both in range and in accuracy. The question of accuracy will be dealt with in a later section. Range requirements are governed mainly by the heights to which information is required and the wind speeds likely to be encountered. A balloon ascending at 6 m./sec. in an average wind of 40 kt. will, if there is no appreciable change in wind direction, be at a slant range of 105 km. by the time it reaches a height of 30 km. Information up to this height is now an everyday need and measurements above 40 km. are likely to be required in the near future. There are at least three ways of improving results in this respect. One is to use faster rates of ascent, but it is difficult to achieve substantial improvement in this way, either by increasing the free lift (since this produces some reduction in bursting height) or by using larger balloons. Another way is to increase the size of the target and so improve the signal returned to the radar. There is, however, a limit to this since it is more difficult to produce larger targets with the necessary precision and to handle them; moreover, they necessitate extra free lift and therefore some reduction in bursting height. The third way is to increase the power of the radar equipment either by using a larger aerial system (and therefore a narrower beam) or by using a more powerful transmitter. There is obviously a limit to decreasing the beam width since if too narrow a beam is used there is difficulty in locating the target. All these possibilities involve greater expense but for economy in routine operation it is generally better to adopt a scheme that increases the performance of the ground station and allows a relatively inexpensive target and balloon to be used.

Other equipment requirements are reliability and ease of operation and maintenance. Generally, a radar should be capable of operating in any part of the world and should therefore possess ruggedness and good mechanical strength and be readily transportable (even though it may be intended for permanent installation in a building). Automatic following facilities are not essential but it is generally considered necessary for the scanning system to be automatic. It must be possible to obtain simultaneous readings of range, azimuth and elevation at intervals of one minute or less.

The choice of wavelength for wind-finding radar is practically restricted to the 10 cm., 5 cm. and 3 cm. bands, since it is for these bands that the special components, such as transmitter valves, have been commercially developed. The longer the wavelength the larger the aerial system required to obtain the same beam width. Generally the diameter of a paraboloid reflector for a directional aerial has to be at least ten times the wavelength to produce a reasonably narrow beam and the longer wavelength also involves the use of larger targets. Mechanical accuracy, particularly when the aerial system is exposed to strong winds, is therefore more difficult to achieve than in the smaller constructions that can be used with the shorter wavelengths. The main objection to the use of 3 cm. for wind measurement is that it is more susceptible to attenuation by raindrops. The 5 cm. wavelength would seem to offer a good compromise but it has not yet been used to any great extent.

Principal types of primary radar in current use for wind-finding.—In the Meteorological Office three types of military or naval radar have been adapted for use as wind-finding equipment. Type 1 is a suitably modified version of the Army gun-laying radar, A.A. No. 3 Mark 2, and has been in use at land stations since about 1943. Type 2 is a Naval search-type of radar, 277P, which is in regular use in Ocean Weather Ships. Type 3 is a modification of the Army radar A.A. No. 3 Mark 7, which differs from the Mark 2 mainly in having a combined transmitter and receiver aerial and in being provided with automatic following gear. There is an American equipment very similar to this known as SCR 584. All these sets are mobile equipments mounted on trailers and they all operate in the 10 cm.

waveband. A non-mobile 10 cm. equipment, known as ME.7, is in use in New Zealand. It was developed from a military equipment that was originally designed as a height finder and its main feature is its very large parabolic aerial reflector. Finally there is the recently developed Decca wind-finding radar, Type WF.1. This differs from the others in operating on the 3 cm. wavelength and although the set is not mounted on a wheeled chassis it is readily transportable. Further details of these equipments are given in the following sections.

4.2. METEOROLOGICAL OFFICE WIND-FINDING RADAR, TYPE 1

4.2.1. The basic equipment

General description.—The Type 1 radar is an adaptation, for the special requirements of wind-finding, of the Army set A.A. No. 3 Mark 2, more familiarly known as the G.L.3. This set was designed as a mobile fire-control instrument for use in anti-aircraft gun laying. As full technical details of the equipment are available at Meteorological Office radar wind stations in the War Office instruction manuals, only a general description is given here, but it is followed by a more detailed account of the Meteorological Office modifications for the wind-finding application.

The equipment provides for the accurate determination of the slant range, azimuth and elevation of a target and for passing this information directly to a predictor. A general view of the equipment is shown in Plate IX. The steel cabin, which is mounted on a four-wheeled trailer that can be jacked up and levelled, contains a cylindrical rotor unit mounted on bearings and capable of being turned about a vertical axis. Above the cabin the rotor supports two paraboloidal aerial systems mounted parallel to each other on a horizontal shaft which can be turned to elevate or depress them. The rotor houses most of the transmitting system and some of the receiving apparatus. The cabin also contains the presentation unit in which the rest of the receiving apparatus and the cathode ray display tubes are housed. At the front of the trailer is a motor alternator set with two synchronous alternators. The whole equipment weighs 9600 kg. (9 tons, 10 cwt.) and requires a power supply of 10 kW. at 230 V. and 50 c./s., which should not fluctuate by more than $\pm$ 2 per cent. of the nominal voltage and $\pm$ 5 per cent. of the nominal frequency. A generator driven by a Diesel engine is available for use where there is no suitable public supply.

The transmitter.—A diagram showing the main electrical units of the equipment is given in Fig. 6. The energy for the transmitter is supplied to the modulator by one of the 420 c./s. alternators and a high-frequency transformer. For the switching operation to enable the modulator to release the energy in the form of pulses, thyratrons were used at first but have been largely superseded by a spark-gap type of modulator. The pulses from the modulator, each of about 25 kV. amplitude and 1 μ sec. duration, are applied to the cathode of a magnetron ultra-high-frequency oscillator. The type of magnetron generally used, namely CV120, oscillates at a frequency of about 3000 Mc./s. and develops, during the pulses, a peak power of 250 to 300 kW. when it is operated in a magnetic field of 1350 gauss. The oscillations are fed through coaxial conductors to the dipole of the transmitting aerial system, which therefore radiates pulses of approximately 1 μ sec. duration at a recurrence frequency of 420 per second. The transmitting dipole is stationary at the focus of its 4 ft. diameter paraboloid reflector. At the same time as the modulator produces a transmitter pulse it also supplies a locking pulse which serves to start the time-bases in the presentation unit.

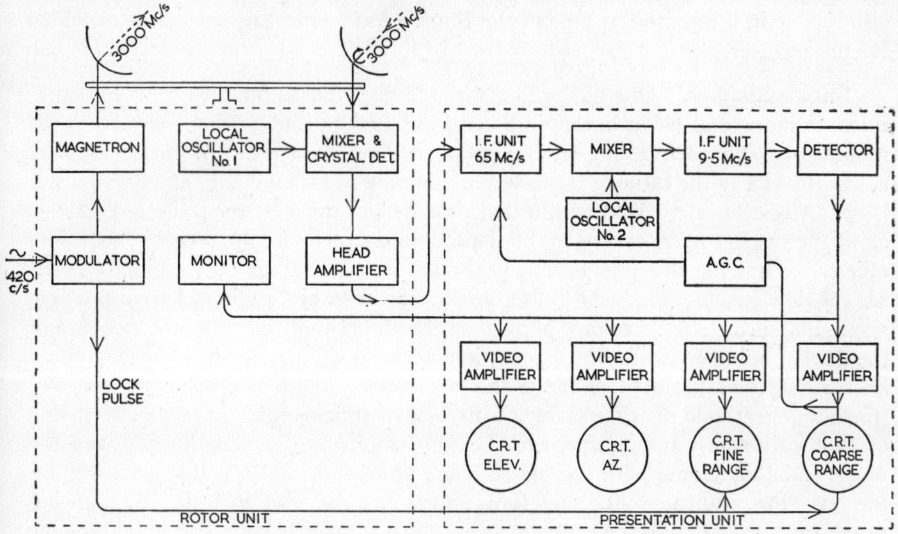

FIG. 6—MAIN ELECTRICAL UNITS OF METEOROLOGICAL OFFICE RADAR, TYPE 1 (MODIFIED A.A. NO. 3, MARK 2)

The receiver.—The receiver aerial system is similar to that of the transmitter except that the dipole is offset from the axis of the paraboloid and can be rotated, by a synchronous motor, about that axis. This allows the receiver beam, which has a width of 6°, to sweep through a cone whose axis is that of the paraboloid and whose angle is about 2°, thus providing for the split-beam method of accurately determining the direction of a target. The rate of rotation of the dipole is 105 c./s. so that four signals are received for each complete turn; the dipole is set in such a way that it is in one of the four cardinal quadrants when a signal is received. A two-phase synchronizing wave of 105 c./s., provided by an alternator driven by the same motor that rotates the dipole, is used to separate the two azimuth signals and mask the elevation signals, while a similar voltage 90° out of phase with the first performs the same function for the elevation display.

The signals received by the rotating dipole are fed through coaxial conductors into a resonant cavity mixer and crystal detector housed in the rotor. By mixing them with the output of a local oscillator a beat frequency of about 65 Mc./s. is produced. This is amplified in what is known as the head amplifier (also housed in the rotor) before being fed through a high-frequency cable to the presentation unit.

The primary function of the presentation unit is to provide control of the aerial mechanism of the rotor unit so as to allow a target to be located and its range, azimuth and elevation to be determined. For this purpose the unit is supplied with the locking pulse from the modulator, the received signal pulses and also with synchronizing waves. The signals coming from the head amplifier at a frequency of about 65 Mc./s. pass into a superheterodyne amplifier and receiver. This incorporates a first intermediate frequency unit, a mixer unit which in conjunction with a second local oscillator converts the signal to a frequency of 9·5 Mc./s., a second intermediate frequency unit and a detector. From the latter the signals are fed through video amplifiers to four cathode ray tubes in the presentation unit and also to one in a monitoring unit in the rotor. Automatic gain control is provided for the receiver.

Range display.—One of the four cathode ray tubes is for the coarse measurement of range and has a linear time-base, initiated by the locking pulse from the modulator. The delineating spot subsequently sweeps steadily from left to right across the face of the cathode ray tube, reaching the right-hand edge about 240 μ sec. later. Any echoes received within that time deflect the spot vertically at positions along the sweep corresponding to the elapsed times and hence the ranges of reflecting objects. Thus the sweep is directly calibrated 0 to 36,000 metres (originally yards, see Section 4.2.2). The echo pulse on the trace can be brightened by a 6 μ sec. signal marker or strobe voltage applied to the cathode of the tube, the position of the marker being controlled by a potentiometer coupled to a handwheel. To obtain an accurate measurement of range the echo pulse is also displayed on a second (fine range) cathode ray tube using a half-cycle of an expanded time-base consisting of 18 c. of a sine-wave voltage, each being of about 13·4 μ sec. duration and the whole train coinciding with the coarse-range time-base. Each half-cycle sweep on the fine-range indicator tube therefore represents a range of 1000 m. By applying to the grid of this tube the same 6 μ sec. signal marker voltage that brightens the echo pulse on the coarse-range sweep, the 1000 m. section that brackets the pulse is displayed on the fine-range tube. The 6 μ sec. strobe potentiometer is then used for adjusting the phase displacement to bring the pulse into coincidence with the reference mark and so give an accurate range measurement.

Range readings may be in error by a fixed amount throughout the scale, or the scale itself may have incorrect intervals. For calibrating the scale intervals a limiting amplifier valve, fed by a 420 c./s. wave, and a crystal-controlled oscillator are switched into the time-base circuit in place of the locking pulse. They produce on the range indicators pips corresponding to range intervals that are accurately known from the crystal frequency, the time-base being fired on the first pip after the onset of the 420 c./s. square wave. In operation the time-base is, of course, used in synchronism with the transmitter and not with the calibration wave. The latter does not check the zero error. This arises from finite delay times in the receiver and from small departures from synchronism between the start of the time-base and the transmitter pulse. It is usually sufficiently constant (until certain critical valves, such as the magnetron, are replaced) to allow of a fixed correction, but a marker target at a known fixed distance may be used for regular checking.

Azimuth and elevation display.—In the display of azimuth and elevation signals on the other two cathode ray tubes in the presentation unit, the received pulses corresponding to opposite quadrants of the conical scanning cycle of the aerial are shown side-by-side and they can be maintained at equal amplitude by controlling the movement of the rotor in azimuth and of the paraboloids in elevation. For this purpose there are, adjacent to each of the indicator tubes, handwheels geared to selsyn generator motors the movements of which are transmitted to receiver selsyns fitted to the rotor unit and to the mounting of the paraboloids. The azimuth and elevation tubes have 4 μ sec. time-bases controlled by the 105 c./s. synchronizing wave from the rotor unit so that brightening pulses are applied to the tubes alternately.

Monitor.—The monitor located in the rotor unit provides, on the screen of a cathode ray tube, visual indication of the operation of the equipment. This is necessary as a check on the efficiency of operation, as an aid to tuning and for the location of faults. The following indications can be obtained on the tube when the equipment is in operation:

(i) The current pulse from the modulator.

(ii) The received signal after amplification, i.e. a similar picture to that shown on the coarse-range tube.

(iii) A sinusoidal calibration wave (1 Mc./s.).

(iv) Any voltage wave form applied through the test terminals provided on the monitor.

The A.A. No. 3 Mark 2 radar, in its original form, did not entirely meet the requirements for wind measurement to high levels and some modifications were essential. In particular, it was necessary to increase the maximum range to well above the 36,000 yd. limit of the original scale and to provide a means of reading accurately the angular information which, for military use, was fed directly to a predictor. Other modifications, the most important of which are a new design of head amplifier and an error sensing unit, have been introduced to improve performance and to facilitate the operation and maintenance of the equipment. Brief descriptions of the principal modifications are given in the following section; full technical details, including installation instructions, are contained in a special (unpublished) series of *Meteorological Office instrument instructions*, with the index letters R.M., on radar maintenance.

4.2.2. Modifications for the Type 1 equipment

Extension of range scale.—The first step in extending the range measurements was a range doubler that was introduced on some of the Army sets which, after this modification, were identified as Mark 2/4. Details are given in the War Office instruction manual provided with the equipment. The principle of the range doubler is to introduce a known time-delay into the locking-pulse line between the rotor and the presentation unit and so cause the range time-bases to be delayed for a time equivalent to 30,000 m., thus allowing measurements to be made from 30,000 to 66,000 m. Provision is made for switching the delayed locking system in and out of circuit. In the " out " position of the switch the output of the delay circuits is disconnected and the range time-bases are triggered in synchronism with the transmitter output, so that range measurements up to 36,000 m. can be made. In the " in " position the delay system is brought into operation and the extended range is then available.

A further modification on the same principle was made by the Meteorological Office to convert the range doubler into a scale trebler. This allows a second time-delay, equivalent to 60,000 m. of range, to be used, thus extending the full scale to 96,000 m. A switch is provided so that either of the delay circuits can be inserted in the time-base locking line or the normal undelayed time-base can be used. To obtain satisfactory signals from the standard types of target at long ranges it is essential for the transmitter to have a high-power magnetron, such as the CV. 120. This necessitates a high power supply for the field magnet and also a polythene casing on the transmitting dipole.

Fine-reading dials.—When the A.A. No. 3 Mark 2, in its original role as a gun-laying radar, fed its information into a predictor only coarse-reading dials for azimuth and elevation were provided on the presentation unit. For wind-finding purposes, fine-reading dials actuated by magslip repeater mechanisms from the rotor and paraboloid mountings are fitted. On these dials one revolution of the pointer corresponds to 10°, so readings of azimuth and elevation to 0·1° are easily made.

Head amplifier, Mark 2.—In order to obtain an improvement in gain and signal-to-noise ratio and so achieve more accurate detection of long-range echoes the original head amplifier has been replaced by a Mark 2 model (Stores Ref., Met. 7471) of more modern design, and which is more readily removable. The unit incorporates a 200 V., 420 c./s. power supply and is therefore self-contained. The amplifier employs a cascode type of circuit, the special feature of which is the use of a combination of a pair of triode valves instead of a pentode. A PCC.84 double-triode valve, specially designed for use in this type of circuit, provides the combination. The first half of the valve is used as an earthed-cathode amplifier and is directly coupled to the second half which is employed as an earthed-grid triode. This arrangement gives excellent stability with a very low noise factor.

A circuit diagram of the amplifier is given in Fig. 7. The input to the grid of the first half of the cascode valve V1, from the mixer unit, is to a tapping on the grid coil L2. The position of this tapping affects the noise factor of the whole stage and has been determined experimentally to give the optimum signal-to-noise ratio. Direct coupling of the two halves of the valve is effected by the anode of the first half being connected to the cathode of the second half through a resistance–capacity bias network R2C6. To extract the amplified signals a tuned circuit with output coupling coil L3 is connected in series with the anode supply to the second half of the valve. The amplifier requires a power supply of 180 V. at 12 m.amp. This is provided by a half-wave rectifier supplied from a 200 V. winding on a mains transformer.

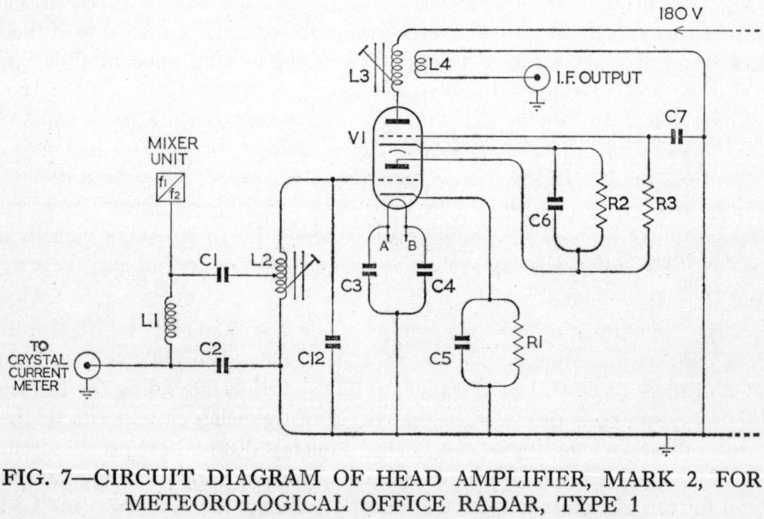

FIG. 7—CIRCUIT DIAGRAM OF HEAD AMPLIFIER, MARK 2, FOR METEOROLOGICAL OFFICE RADAR, TYPE 1

Error sensing unit.—The method of keeping the aerial system correctly aligned on a target by matching pairs of " pips " on the azimuth and elevation indicator tubes becomes difficult when signals are fading or fluttering, especially at the longer ranges. An improved method of obtaining correct alignment is provided by an error sensing unit (Mark 1, Stores Ref., Met. 7435). This instrument separates successively received pulses into batches corresponding to azimuth, left and right, and elevation, up and down, and produces a direct current proportional to the mean amplitude in each channel over a period of time. Corresponding pairs of outputs are fed into a centre-zero meter which indicates zero if the outputs are equal but deflects one way or the other if the beam is off target. The unit also incorporates a channel-switched automatic gain control system which applies automatic gain control independently to the azimuth and elevation signals, thus helping to remove flutter and to minimize the effects of fading. Full technical details of the unit and its installation are given in *Meteorological Office instrument instructions* R.M. 33 and 36 but a brief description is given here, illustrated by a block diagram in Fig. 8.

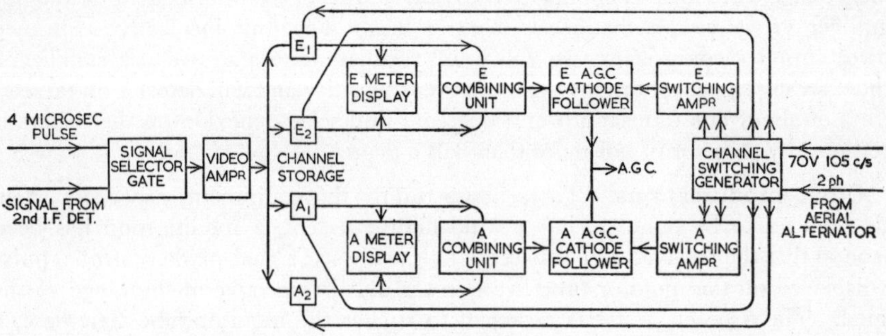

FIG. 8—BLOCK DIAGRAM OF ERROR SENSING UNIT FOR METEOROLOGICAL OFFICE RADAR, TYPE 1

E_1, E_2, A_1 and A_2 are capacitor–triode circuits for elevation and azimuth.

The signal from the 9·5 Mc./s. detector in the radar receiver is fed to a gate valve which is also supplied with the 4 μ sec. strobe pulse. A pedestal pulse is thereby produced, upon which is superimposed the desired echo. After the removal of the pedestal pulse by a biassed diode the signal pulse is applied to the grid of a video amplifier. The anode load of the latter is a pulse transformer of ratio unity, the secondary of which is tuned to resonate a little above 420 c./s. by a low-loss capacitor. A considerable lengthening of the pulse is thus produced in this part of the circuit. The return path of the secondary is through a triode valve. There are four of these capacitor–valve circuits in parallel in the secondary, to correspond with the four channels it is required to separate, and all except one of the triodes are maintained in a non-conducting state. Switching voltages, which are applied to the grids of the triodes, are derived in quadrature from two centre-tapped transformers whose primaries are connected across the two-phase output of the aerial alternator. Each of the four low-loss capacitors with the associated leakage resistance integrates the pulses received from one of the four cardinal positions of the receiving aerial with a time-constant of a few seconds, and thus acquires a charge approximately proportional to the mean amplitude of the pulses over this time interval. Hence, when a centre-zero voltmeter is connected across corresponding pairs, a difference in the accumulated voltages will show as a deflexion to left or right indicating that the aerial system is off target and that appropriate adjustment of the handwheels is needed to bring it on to the target and the indicator to zero.

The mean amplitude of the voltage developed across corresponding pairs of capacitors, measured with respect to earth, is proportional, within limits, to the amplitude of the signal; hence it may be used for the automatic control of the intermediate-frequency amplification. Since two mean voltages are available, one from the elevation channel and the other from the azimuth channel, it is possible, by switching from one to the other, to control each set of signals independently. This represents a considerable advance on the former automatic gain control channel and helps to ensure that the signals are displayed for the maximum time. The combining and switching of these two channels is achieved by feeding the control voltages to the grids of two cathode followers which have a common cathode load, and the screens of which are square-wave pulsed by the transformers that switch the capacitor–valve circuits. In this way each cathode follower produces an output alternately and the voltage at the common cathode is a series of square waves of amplitudes alternately proportional to the azimuth and elevation signal strengths. This voltage wave form is applied to the control grids of the intermediate-frequency amplifier valves as automatic gain control bias. The unit incorporates its own power supplies comprising two full-wave rectifier units and two gas stabilizers. Under strong signal conditions the unit is very sensitive and will detect " off target " errors of about one thousandth of a degree. Full scale deflexion on the meters is produced by an error of not more than half a degree.

Other modifications.—To facilitate tuning the equipment, especially when this requires to be done during a wind-finding ascent, a modification has been made so that the received signal selected by the 6 μ sec. signal-marker (strobe) pulse is displayed on the monitor tube in the rotor unit irrespective of the range of the signal. The 6 μ sec. pulse is arranged to trigger the monitor tube time-base, a switch being provided to allow normal operation of the time-base when required.

Another modification of the equipment became necessary when it was decided to adopt metres instead of yards as the unit for measuring range so as to facilitate the reporting of heights in metric units. This change was made by replacing the crystal controlling the valve oscillator used for calibrating the time-bases. In the original equipment a quartz crystal giving a frequency of 164 kc./s. was used so that the time for one cycle was that taken by an echo from 1000 yd. In the modification this crystal was replaced by one (Stores Ref., Met. 7530) having a frequency of 149·896 kc./s. Taking 299,792 km./sec. as the velocity of electromagnetic radiation this gives one cycle as equivalent to one kilometre of range.

In order to ensure a voltage supply stable to within $\pm$ 1 per cent., an automatic voltage regulator of type BMVR 7000A and total capacity 7 kV. amp. (Stores Ref., Met. 9463) has been installed in Meteorological Office equipment. This controls all supplies to the apparatus in the cabin except the air-conditioning unit and service switch plug sockets. Additional voltage regulation is provided for the supply to the presentation unit by the fitting of a regulator of type BAVR 1000 (Stores Ref., Met. 9464) in place of the Variac auto-transformer normally fitted. This regulator minimizes fluctuations of periods down to about 20 m. sec.

The remaining modifications which have been introduced on the Type 1 radar are of a minor nature. They include the provision of an auxiliary heater to sets fitted with a thyratron modulator, a filament switch for removing heater supply to a magnetron that appears to be unstable (thereby increasing the useful life), and a means of illuminating the range dial by a perspex annulus which reflects light internally from four lamps inserted in the outer edge.

4.2.3. Installation and maintenance

Installation.—The equipment should be installed on high ground with the horizon as free from obstructions as possible. Metal objects such as corrugated iron roofs of farm buildings or wireless masts anywhere within the horizon will cause confusion of signals, but a single isolated object a few miles away may be useful as a reference marker. There should be no extensive obstructions subtending an angle of elevation exceeding 3° at the observation point, otherwise targets may become obscured in strong winds. A symmetrical hill with a downward slope of about 6° for a radius of 400 m. in a hollow surrounded by hills rising to 1° or 2° elevation, affords a good site since it avoids ground echoes beyond short range. A flat site clear of obstructions is also suitable. The site should be provided with a firm foundation for the equipment to stand on without sinking or changing its level. This should be either a concrete floor or concrete blocks firmly set in the ground. The equipment is levelled by means of its three jacks, the feet of which rest on plates placed on the concrete.

Orientation of radio and optical axes.—The following conditions must be fulfilled in order to ensure accuracy in azimuth and elevation measurements:

(i) The axis about which the aerials are rotated in order to measure azimuth must be vertical (to within ± 1 min. of arc). This is secured by means of a spirit-level on the rotor head.

(ii) The axis about which the aerials are rotated in order to measure elevation must be horizontal (to within ± 3 min.). A spirit-level fitment is provided for securing this.

(iii) The optical axis of the telescope fitted to the aerial system must be perpendicular to the axis of elevation (to within ± 3 min.). This is checked by means of an auxiliary telescope and involves the selection of a reference object at zero elevation.

(iv) The radio axis of the receiver aerial system must be parallel to the optical axis of the telescope (to within ± 3 min.).

(v) The axis of the transmitter aerial system must be parallel to the optical axis (to within ± 15 min.).

The detailed procedures for adjusting the equipment to fulfil this condition and for checking the accuracy of orientation are given in *Meteorological Office instrument instructions* R.M. 12 and 13 and in the *Handbook of radio-sounding technique*[46].

Operation, adjustment and maintenance.—Full details of the procedures for starting up the equipment, tuning the transmitter and receiver and setting up the time-base and the presentation unit controls are given in the War Office instruction manual on radar equipment. Tuning of the equipment involves the adjustment of the various controls of the transmitter and receiver so as to obtain (*a*) the transmission of a steady high-frequency pulse at the required frequency and of maximum power and satisfactory shape and (*b*) the reception of a reflected signal of maximum signal-to-noise ratio, as steady as possible and of satisfactory shape. The overall performance is best measured by using the echo from an airborne target of the type that is normally used in the routine work.

Instructions for maintenance and servicing are also given in the manual referred to above. They are supplemented by *Meteorological Office instrument instructions* in the R.M. series and, in particular, R.M. 2 gives the detailed schedule of weekly and monthly maintenance tasks which should be undertaken in order to keep a Type 1 radar in good condition. It cannot be emphasized too strongly that the operational performance depends very much on the care and skill with which the equipment is maintained. In addition to the usual items of electrical equipment, such as meters and measuring bridges that are used in testing various parts of the equipment, three special items are provided for use in maintaining and checking the performance of the Type 1 radar. They are a signal delaying resonator, a signal generator and a field-strength measuring set.

Signal delaying resonator, H.F. No. 1., Mark 1 or 2 (Stores Ref., Met. 3560).—The main purpose of the resonator is to provide a reliable signal for use in tuning and checking the transmitter and receiver circuits of the equipment. The resonator, which is illustrated in Plate IX, is a tuned circuit of the resonant cavity type with a very high Q (magnification factor) and, consequently, a very small decrement. It is provided with a radiator which can be mounted on the cabin roof where it will pick up signals from the main transmitter. The signals are fed through a radio-frequency lead to the resonator. Tuning to the transmitter frequency is done by a micrometer adjustment, using a micro-ammeter as an indicator. The cavity continues to resonate after the end of the transmitter pulse and thus continues to re-radiate energy. If the receiver circuits are tuned to the transmitter frequency and are operating properly an echo will be observed on the range time-base. Thus, by observing the effect on the range tube, various tuning adjustments can be made to the transmitter and receiver circuits. Also, different crystals may be tried in the receiver mixer stage and the relative efficiencies compared in terms of range. By taking micro-ammeter readings at several settings of the resonator micrometer throughout the frequency band covered by the transmitter, a rough check can be made of the frequency spectrum of the magnetron and of the general stability. Further technical details of the instrument are given in the War Office manual on radar equipment.

Signal generator No. 6A. (Stores Ref., Met. 3563).—This instrument, which is illustrated in Plate X, is designed to produce a signal of known frequency, modulation and amplitude. It is used primarily for obtaining an accurately reproducible, though not absolute, measurement of the signal-to-noise performance of the receiver. This measurement is carried out by tuning the signal generator accurately to the receiver at a convenient spot frequency and noting the power input required to produce a given signal-to-noise ratio at the video output stage. The receiver design is such that the performance at any spot frequency within the normal tuning range may be taken as a reliable guide to the performance over the whole of that range.

A velocity-modulated reflection (Klystron) oscillator is used to generate a signal at a wavelength of 10·7 cm. ± 0·1 cm., with square-wave modulation at 1000 c./s. The signal feeds a piston-type attenuator, a glow-lamp being used as the exciting source and visual monitor of the power. The output is applied to the mixer of the radar set by means of a length of attenuating cable and the video output from the receiver is taken back to the signal generator and displayed on a built-in cathode ray tube monitor, the time-base of which is locked to the modulation frequency. Adjustment is then made to the attenuator control until a 1 : 1 signal-to-noise ratio is obtained. The reading of the attenuator then indicates the performance of the radar receiver at its noise level. Technical details of the signal generator are given in the War Office instruction manual and its method of use in *Meteorological Office instrument instructions* R.M. 5 and 29. It should be emphasized that the instrument is a very delicate piece of equipment. Any damage caused to the radio-frequency section or the output cables necessitates re-calibration, which is an expensive procedure.

Field strength measuring set, R.F. No. 3 (Stores Ref., Met. 3562).—This equipment provides a convenient means of making a frequent check of the output

GENERAL VIEW OF METEOROLOGICAL OFFICE RADAR, TYPE 1
(MODIFIED A.A. NO. 3 MARK 2)

SIGNAL DELAYING RESONATOR, H.F. NO. 1, FOR TUNING AND CHECKING PERFORMANCE OF METEOROLOGICAL OFFICE RADAR, TYPE 1

Plate X

SIGNAL GENERATOR NO. 6A, FOR MEASURING RECEIVER PERFORMANCE OF METEOROLOGICAL OFFICE RADAR, TYPE 1

GENERAL VIEW OF METEOROLOGICAL OFFICE RADAR TYPE 3
(MODIFIED A.A. NO. 3 MARK 7)

Plate XII

By courtesy of Barratt's Photo Press Ltd.

AERIAL SYSTEM OF METEOROLOGICAL OFFICE RADAR, TYPE 2 (MODIFIED 277 P)

of the transmitter. It consists of a dipole aerial with a filament connecting the two halves. The filament forms the heater of a thermocouple and the whole is enclosed in an evacuated glass envelope and connected to a moving-coil meter which is calibrated in arbitrary field strength units. The apparatus is mounted on a mast about 3 m. high, as shown in Fig. 9, and is set up about 30 m. from the radar set. As the d.c. voltage produced by the thermocouple depends on the temperature of the junction, and as this depends on the power used in heating it, the meter indicates the power absorbed by the aerial and therefore the field strength. The relation of current to field strength is not linear, however, so the scale is divided into five equal divisions which are arbitrarily taken as units of field strength. For checking the output of the radar transmitter the readings are taken with the transmitting paraboloid beamed on to the meter. With the measuring set at 33 m. from the radar the output should not be less than 3 units of field strength.

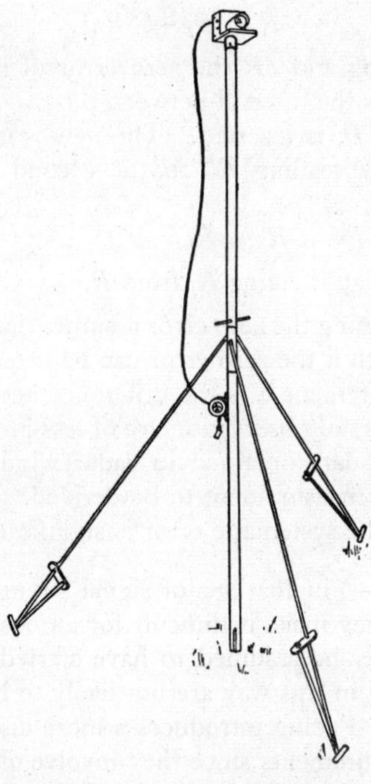

FIG. 9—FIELD-STRENGTH MEASURING SET, R.F. NO. 3, FOR CHECKING TRANSMITTER OUTPUT OF A METEOROLOGICAL OFFICE RADAR, TYPE 1

The measuring set can also be used for tuning the pair of stubs that are provided for matching the impedance of the magnetron with that of the feeder to the aerial. The stubs are adjusted until a transmitted pulse of maximum amplitude is obtained. Further details of the measuring set and its method of use are given in the War Office manual on radar equipment.

4.2.4. Errors

Systematic errors.—If the equipment is properly levelled to ensure that its rotor axis is vertical and its elevation axis horizontal and if the correct procedure (see Section 4.2.3) is carried out for the alignment of the optical and radar axes the systematic errors in azimuth and elevation measurements should not exceed $\pm$ 3 min. of arc. Systematic errors in the scale intervals of the range indicators, after these have been calibrated by means of the crystal oscillator provided for this purpose, should not exceed $\pm$ 10 m. Zero error in the range scale, which may arise from finite delay times in the receivers and from small departures from synchronism between the times of start of the time-base and the transmitter pulse, should generally remain reasonably constant. It can be determined by measurements on a fixed target at an accurately known range, and allowed for by an appropriate correction. A more convenient method is to use an electronic zero-error calibrator, as described by Allwood *et alii*[47]. Briefly this involves the measurement of the apparent range of an artificial signal generated by the calibrator and the comparison of this with the true figure. Two such signals are needed: the earlier is set to coincide with the transmitter pulse and the second is set to record the range reading R on the time-base. Then,

$$R = R_t + \delta R,$$

where R_t is the true reading and δR the zero error of the time-base. The true range of the second signal is the interval between the two pulses. To measure this interval an arbitrary delay, D, is inserted. The new reading R' to the first pulse is $D + \delta R$, while the new reading R'' to the second pulse is $D + R_t + \delta R$. Subtraction gives

$$R'' - R' = R_t$$

and δR is then obtained by subtracting R_t from R.

This method of determining the zero error assumes that there are no non-linear errors in the time-base; with it the zero error can be determined to within $\pm$ 1 m. Since in upper wind measurement it is the difference between successive readings that is used, systematic errors of observation are of less importance for this purpose than random errors. The data obtained in radar wind-finding enable also the mean altitude of each wind measurement to be derived. When this is done, as in a sounding of wind alone, the systematic error is significant.

Observational errors.—Fluctuations of signal strength are one of the causes of observational errors. They make it difficult for an observer to decide at which point in range the echo may be assumed to have started. Unless fading is very bad, however, errors arising in this way are not likely to be serious, the maximum error being about $\pm$ 30 m. Fading introduces a more disturbing element into the azimuth and elevation measurements since they involve observations of amplitude. Errors in these measurements can be reduced by a quick-acting automatic gain control circuit and the use of the error sensing unit should minimize these errors.

Trials have shown that with a well maintained set the standard deviation of the random errors is less than 15 m. in range and $0{\cdot}1°$ in azimuth and elevation so long as the signal is fairly steady and the signal-to-noise ratio is better than about 3 : 1. At long ranges, when the signal deteriorates the errors increase.

The effect of the errors quoted above on wind measurements can be calculated by means of the formula (i) in Section 2.3.2. Some results of such calculations were reported by Bannon[48]. They included a table giving the standard deviation of vector wind error as a function of height and mean wind; this table indicated that the standard deviation in determining the mean wind in a layer approximately 350 m. thick is not greater than 2 kt. up to a height of 3 km. and seldom exceeds 3 kt. at 6 km. or 5 kt. at 18 km.

4.3. METEOROLOGICAL OFFICE WIND-FINDING RADAR, TYPE 3

4.3.1. The basic equipment

The Type 3 radar is an adaptation of the mobile anti-aircraft fire-control equipment, A.A. No. 3 Mark 7, sometimes known as the G.L.7, a general view of which is reproduced in Plate XI. It is considerably lighter than the Mark 2, the total weight being a little over 5000 kg. (5 tons). Having automatic following, it is designed to provide continuous information on range, azimuth and elevation in the form of magslip voltages for feeding directly into a predictor and also for display on repeater dials. The cabin, on a four-wheeled trailer fitted with a three-point jacking system, houses a presentation unit, containing the radar circuits, and a mount containing the gearing for rotating the aerial system in azimuth and elevation, together with the data transmission system. In front of the cabin two alternators and two power amplifiers (amplidynes), with their driving motors, and also a refrigeration unit are mounted under covers. The combined transmitting and receiving aerial system, with a 5 ft. diameter paraboloid, is carried on the upper part of the mount.

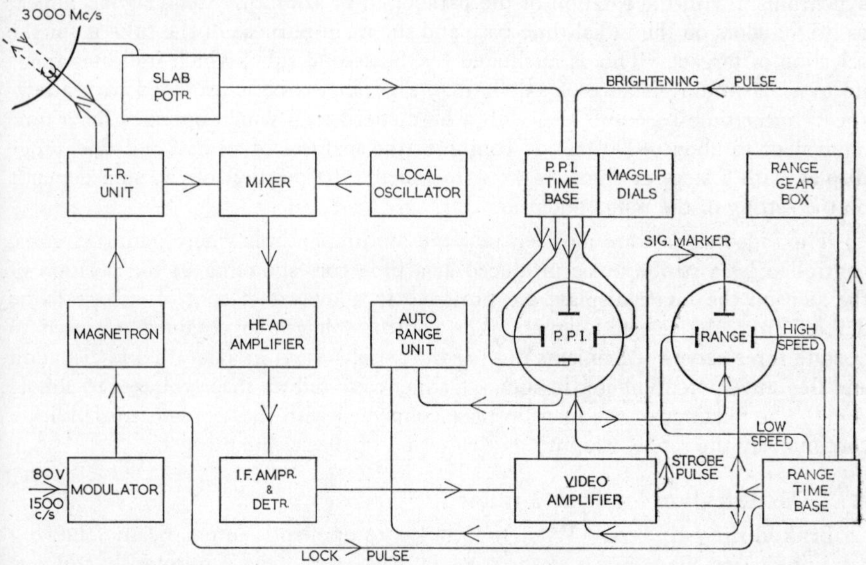

FIG. 10—BLOCK DIAGRAM OF MAIN ELECTRICAL UNITS OF METEOROLOGICAL OFFICE RADAR, TYPE 3 (MODIFIED A.A. NO. 3 MARK 7)

Full technical details of the equipment are given in the War Office instruction manual provided with the equipment; the main electrical units are shown in the block diagram in Fig. 10. The modulator, which is a single spark-gap type, is supplied with a 1500 c./s. input from one of the alternators and produces high-voltage d.c. pulses of 0·55 μ sec. duration 1500 times per second. These are fed to a cavity magnetron of type CV160 which converts them into radio-frequency pulses with oscillations at a frequency of about 3000 Mc./s. and a peak power of 200 kW. After passing through a T.R. switch (to prevent large amounts of radio-frequency power reaching the receiver) the output from the magnetron is fed asymmetrically into a half-wave dipole aerial mounted symmetrically at the focus of the paraboloid. The dipole is rotated by the aerial motor at 33 revolutions per second and since it is fed asymmetrically the radiation from the aerial system diverges slightly from the axis of the paraboloid. The polar diagram therefore sweeps the surface of a cone, the axis of which coincides with that of the paraboloid. Echoes from targets displaced from this axis will thus be modulated at 33 c./s.

At the end of each transmitter pulse the T.R. switch allows received pulses to pass to the receiver mixer circuit into which oscillations at a frequency slightly below that of the magnetron are fed from a reflex klystron local oscillator. The output from the mixer is an intermediate frequency of 60 Mc./s. and automatic control of this frequency is provided by a subsidiary mixer circuit. The I.F. signals are amplified in two head-amplifier stages and then passed to the I.F. amplifier and detector unit, the gain of which can be controlled manually, or automatically from circuits associated with the auto-range unit. From the I.F. unit the video signals are fed to the video amplifiers and thence to two cathode ray tubes in the display unit and to the auto-range unit.

One of the tubes has a plan-position indicator display, i.e. it indicates the plan position of all echo-producing targets within the range of the set. The time-base is radial and is produced electrostatically by a slab potentiometer which rotates in synchronism with the rotation of the paraboloid in azimuth. Echo signals appear as bright spots on the radial time-base and the main purpose of the tube is for the selection of targets. This is facilitated by the second tube, which indicates range up to a maximum distance of 44,000 m. The range tube is provided with a low-speed linear time-base and also with a high-speed scan which operates for a time equivalent to about 4000 m. It combines the features of coarse and fine range display with a step, or signal marker, included; the position of the step depends on the setting of the range potentiometer.

The video signals are also fed into the auto-range unit where gating circuits, controlled by a strobe pulse produced at a time corresponding to the position of the step on the range display, are provided to allow a d.c. control voltage to be derived from the selected signal. The control voltages so obtained are used to operate three servo-mechanisms driving the aerial system in azimuth and elevation and the range potentiometer in such a manner as to reduce these voltages to almost zero. The readings of the magslip dials connected with the servo-motors indicate continuously the range, azimuth and elevation of the selected target.

4.3.2. Modification for Type 3 radar

Braked magslip box.—When the Mark 7 equipment is automatically following a moving target the pointers of the magslip dials are moving continuously and it is impracticable, therefore, for one operator to take the readings of range, azimuth and elevation simultaneously. In order to facilitate this a method of braking the magslip system has been developed. It consists of the interruption of the reference excitation current and the application, through a relay mechanism, of rubber pads to large diameter extensions of spindles of the magslip repeaters. This permits the pointers to be arrested almost instantaneously (the lag is of the order of a few milliseconds) and to be held in position as long as necessary, thus obviating the risk of erroneous readings, due to inertia of the moving parts of the magslips, which can occur if no braking is applied when the excitation current is switched off. The braked magslip box and its control key can be used remotely from the equipment.

Range-scale trebler.—Extension of the range scale has been obtained by using the added time-delay method but for various reasons the unit is more complicated than that devised for the Type 1 radar. Greater stability is necessary because any drift is equivalent, in the automatic ranging circuits, to an algebraic addition to the rate of change of range of the target. For upper wind measurements the servo-mechanisms are adjusted to work with relatively low range-rates, and drift in the scale trebler unit may increase the rates to such an extent that the set ceases to follow the target. For the same reason the trebler unit must not exhibit any small, sudden variations in its delay time as these would be equivalent to short bursts of very high range-rates.

The scale trebler which has been devised is semi-automatic in operation. When the echo is approaching a range near the maximum of the scale a manually operated key inserts a delay equivalent to 30,000 m. and the range potentiometer controlling the strobe spot is automatically wound back by this amount. During this operation, which takes about ten seconds, the rates of follow in azimuth and elevation existing immediately before the scale change are maintained by a " follow through " circuit. As soon as the range potentiometer has wound back, automatic following is resumed.

4.3.3. Operation and maintenance

Detailed technical instructions for the operation and maintenance of the A.A. No. 3 Mark 7 equipment are given in the War Office instruction manual provided. These instructions cover the installation of the equipment on an observing site, and a complete physical, mechanical and electrical check, which should be made if the equipment has been out of use for more than a few weeks. The complete setting-up of all controls and adjustments is dealt with and is followed by instructions for setting-up for daily operation and for normal switching on and off. The maintenance instructions deal with the equipment unit by unit, in each case under the general headings:

(*a*) Removal of the unit.
(*b*) Dismantling and replacement of components.
(*c*) Servicing and repair.
(*d*) Special faults, where applicable.

Schedules of regular maintenance tasks are included. Although the Type 3 equipment is considerably more complicated than the Type 1, maintenance is made easier by the detachable unit form of construction and by the provision of more metering and monitoring points.

In using the equipment for upper wind measurement the following procedure has been found to be convenient. The aerial system is put on target by using the automatic scan in azimuth and manual control of elevation. The signals are fed into the range display through a delay line equivalent to 1050 m. in range. This enables the echo signal to be observed through the transmitter signal. With this procedure the echo can normally be located by the time the target has reached a range of about 550 m. and the aerial can be " locked-on " at a range of about 870 m. As soon as the set is " locked-on " the delay line is switched out of use so that correct ranges are indicated by the range magslip dials. The range display tube is used in locating the target since it is difficult to obtain a satisfactory combination of control settings for the plan-position indicator display to show the target clearly at short ranges.

4.3.4. Errors of observation

Comparisons of simultaneous measurements with Type 1 and Type 3 have been made and reported on by Harrison[49] and Else[50]. The results indicate that the accuracy of the Type 3 equipment is about the same as that of the Type 1, the estimates of the errors of which are given in Section 4.2.4.

4.4. METEOROLOGICAL OFFICE WIND-FINDING RADAR, TYPE 2

4.4.1. Description

The Type 2 equipment, which is used for upper wind measurements in Meteorological Office ocean weather ships, is basically a search type of radar, known as 277P, designed for Naval use. Technical details of this set and its maintenance are given in an Admiralty instruction manual provided with the set. The aerial system, shown in Plate XII, has a single paraboloid reflector, 1·4 m. in diameter, made of wire mesh and fed by a waveguide. The half-power beam width is about 3°. The aerial is mounted on a pedestal which allows it to be rotated through a sector − 5° to + 75°. Power drives are provided for the movement in azimuth and also for stabilizing gear. The latter is designed to stabilize the aerial in respect of azimuth changes arising from variations in the ship's heading, and also in respect of changes in elevation zero taking place in the vertical plane through the beam. Thus, the effects of yaw and of components of roll and pitch in the direction of the beam should be eliminated, but not the effects of components perpendicular to the beam. The aerial on its pedestal is sited on the tripod foremast of the weather ship and the power drives are controlled from a cabin under the ship's bridge, housing the transmitter, the receiver and the presentation unit.

The block diagram in Fig. 11 shows the main units of the radar and their connexions. The magnetron transmitter radiates on a frequency very close to 3000 Mc./s., with alternative pulse widths of 0·6 and 2 μ sec. and a recurrence frequency of 500 pulses per second; the peak power is about 500 kW. An intermediate frequency of 13·5 Mc./s. is produced by the crystal mixer and the signal is amplified by a two-stage I.F. unit in a head amplifier and a four-stage unit in the receiver.

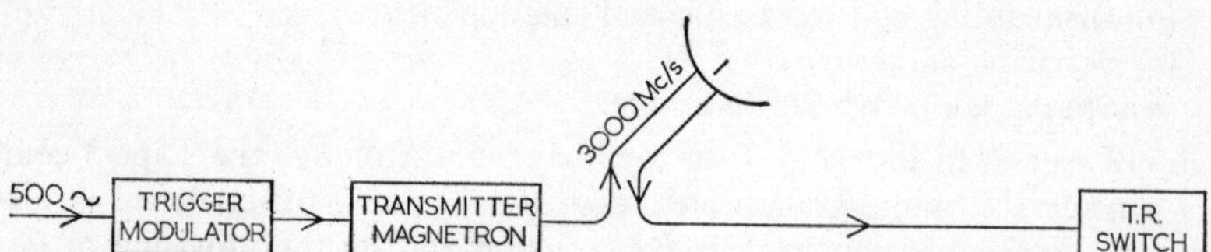

FIG. 11—BLOCK DIAGRAM OF MAIN ELECTRICAL UNITS OF METEOROLOGICAL OFFICE RADAR, TYPE 2 (MODIFIED 277P)

The indicator unit has three cathode ray tube displays. One of these is a plan-position indicator display indicating azimuth and approximate slant range. The circular sweep is produced by mechanical rotation of the deflexion coils of the P.P.I. tube in synchronization with the aerial scan, relative to the heading of the ship. The scanning is motor-driven but the rate of sweep is manually controlled from zero to 90° per second in either direction. Two scales, graduated in whole degrees, are provided for the azimuth tube; one is fixed and the other is controlled by the ship's compass gyro and thus allows for the ship's heading. Elevation and approximate slant range are indicated in polar co-ordinates on another tube; the angular sweep is geared to indicate three times the aerial scan in elevation. As the latter is manually controlled the rate of sweep is low. It is not practicable to scan in azimuth and elevation simultaneously. Cursors are provided for setting on the echo signals but the spread of the signals on the trace is often such as to make accurate setting impossible. The range scales on the azimuth and elevation displays can be switched to cover ranges up to 14, 70 and 140 km. The third tube that is provided is for the more accurate indication of range on an A-type of display. A brightened strobe marker is set against the leading edge of the echo signal, the position of the marker being controlled by a hand-operated dial on which range is read. Range scales can be selected by switch independently of the other displays.

4.4.2. Meteorological Office modification

Delayed synchronization and signal generator.—The purpose of this unit is to enable echoes at long ranges to be located with the same accuracy as those at shorter ranges. It is inserted in the synchronizing line between the modulator and the indicator unit, as shown in Fig. 11. A three-way switch enables the synchronizing pulses to be passed directly to the indicator unit or to be delayed by times equivalent to 30,000 and 60,000 m. of range respectively. The effect is to provide a very long time-base, one third of which is actually displayed on the range tube. In the three positions of the switch, therefore, the ranges covered are 0–30, 30–60 and 60–90 thousands of metres.

In order to ensure accurate measurement in the higher ranges it is necessary to set up the synchronizing delays very accurately. This could be done by reference to suitable permanent echoes at ranges of about 34,000 and 64,000 m. but as such facilities are unlikely to be available at sea, provision is made for the delayed synchronization unit to generate an artificial signal, equivalent to suitable ranges, for use as a reference when the delay circuits are being adjusted.

Range expansion unit.—The purpose of this unit is to provide a high-speed time-base, equivalent to approximately 1000 m. in range, locked to the range strobe of the indicator unit. This permits display of the echo signal on an expanded range scale, with consequent improvement in accuracy.

The range strobe pulse is fed into the unit and the leading edge initiates a linear time-base of 6 μ sec. duration. This is amplified and applied to the X plates of a small cathode ray rube. A brightening pulse applied to the cathode of the tube brightens the trace during the period of the scan. Video signals are also fed into the unit and, after amplification, are applied to one Y plate of the cathode ray tube. The other Y plate allows calibration signals from a standard oscillator to be displayed.

When the strobe marker is set on the required echo signal on the range display tube the same signal will be displayed on the range expansion unit. The accurate range of the echo is then obtained by adjusting the range handwheel so that the leading edge of the echo coincides with the hairline provided on the small cathode ray tube, the dial geared to the handwheel giving the range reading.

4.4.3. Performance and accuracy

The maximum range to which satisfactory signals can be obtained with the Type 2 radar from a corner reflector of 1·37 m. hypotenuse is not less than 100 km. and in this respect the equipment is better than Types 1 and 3. It is, however, much less accurate. There are three main reasons for this. In the first place, owing to incomplete stabilization of the aerial system the effect of the movement of the ship is not entirely eliminated. Secondly, the aerial scanning system does not give such high precision as the split-beam technique. Thirdly, the presentation, while it may suffice for search purposes, is not adequate for precision measurements. On the azimuth and elevation displays the echoes are spread out into arcs of the order of 10° and the process of bisecting these with the cursor is not one which is conducive to high accuracy. Moreover, since sweeping in elevation cannot be started until the azimuth setting is completed, it is impracticable to obtain simultaneous observations of azimuth and elevation even if two operators are available. Also signal fluctuations (e.g. due to rotation of swinging of the target) and asymmetrical movements of the ship about the vertical during the relatively long period of scan, may alter the centre of gravity of the arc displayed.

Estimates of the errors of observations of Type 2 radars in ocean weather ships have been made by Harrison[51] from two series of trials. In these trials direct comparisons were made of the readings obtained by two ships' radars following the same target at sea. Between the first series in 1949 and the second in 1955 the radars had been fitted with the delayed synchronization and signal generator and the range expansion unit. This probably accounted for a considerable improvement in the range accuracy as shown by the systematic differences of range readings between the two sets, amounting to more than 1800 m. in the first series and not more than about 450 m. in the second series. There was no corresponding change in the systematic differences in azimuth and elevation readings, the maximum differences being about 9° and 4° respectively.

The scatter of the individual differences in an ascent showed some long-period fluctuations, so the random component of the observational error was derived from the changes between successive differences in readings, which are of more direct consequence to the accuracy of wind measurement. In both series of trials the standard deviation of the azimuth and elevation errors so derived were 0·5° and 0·3° respectively. The standard deviation for range errors was 110 m. in the earlier comparisons and 60 m. in the later ones. Comparisons were also made between the vector winds derived from the observations. The results showed the mean vector differences to be 7·4 kt. and 6·8 kt. in the first and second series of trials. In the latter, 46 per cent. of the vector differences were less than 4·5 kt. and 89 per cent. less than 14·5 kt.

4.5. DECCA WIND-FINDING RADAR, TYPE W.F.1

4.5.1. Description

This equipment, exterior and interior views of which are reproduced in Plate XIII, is a transportable set with a paraboloid aerial system supported on a framework in front of a cabin, both being carried on a turntable mounted on an adjustable tripod base. The total weight is 2505 kg. (about 2·5 tons). For a normal upper wind sounding two operators are required and both are accommodated in the cabin, which houses all the units of the equipment except the aerial.

The set has a hard valve pulse modulator operating on a frequency of 9375 ± 30 Mc./s., with a nominal peak power of 20 kW., a pulse width of 0·5 or 0·25 μ sec. and a recurrence frequency of 1000 pulses per second. The common transmitter and receiver aerial system, with a paraboloid reflector of 2·5 m. diameter, is fed by a rotating waveguide offset from the focus. A beam of 1° width (to half-power points)

By courtesy of Decca Radar Ltd.

EXTERIOR VIEW OF DECCA WIND-FINDING RADAR, TYPE W.F.1

By courtesy of Decca Radar Ltd.

INTERIOR OF CABIN OF DECCA WIND-FINDING RADAR, TYPE W.F.1

Plate XIV

By courtesy of National Publicity Studios, New Zealand

AERIAL SYSTEM OF NEW ZEALAND WIND-FINDING RADAR, TYPE M.E.7

By courtesy of National Publicity Studios, New Zealand

NEW ZEALAND WIND-FINDING RADAR, TYPE M.E.7

Plate XVI

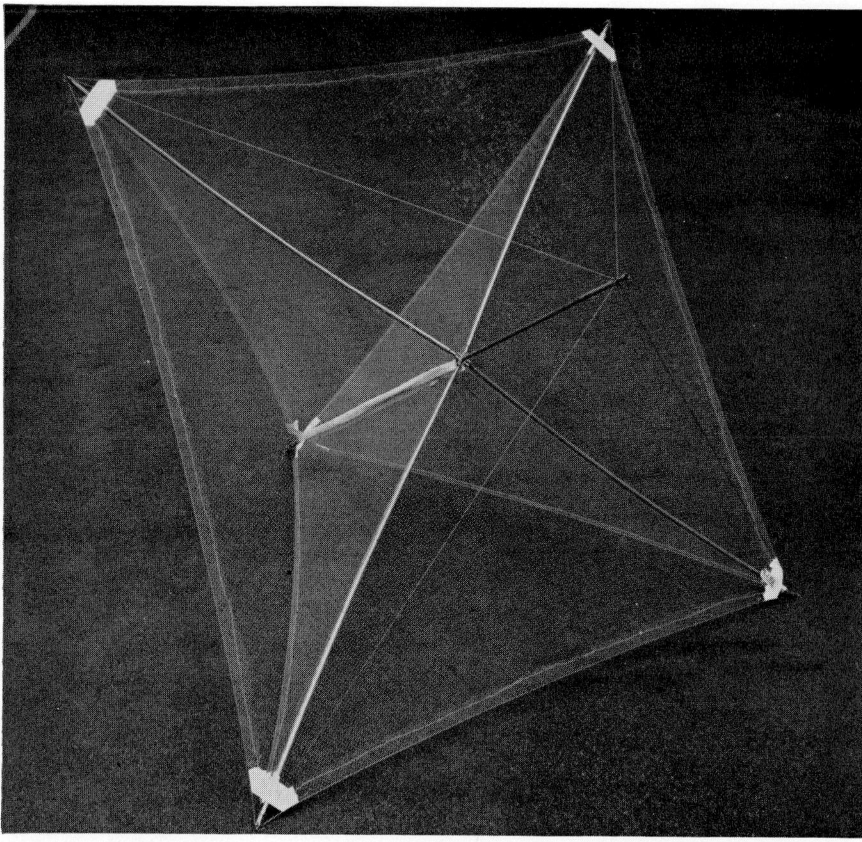

METEOROLOGICAL OFFICE CORNER REFLECTOR, MARK 3

is produced and the rotating feed spins the axis of the beam in a cone of 1° apex angle. Movement of the aerial system in azimuth and elevation is manually controlled through direct gearing from a pair of handwheels which can be turned by one operator. The angles are indicated on dials which can be read to 0·1°.

The received signal, after conversion to an intermediate frequency of 30 Mc./s., is displayed on a single cathode ray tube with I-scope indication. In this type of display the echo is shown as a circle or arc of a circle, depending on whether the aerial is accurately set " on target " or whether it is slightly " off target ". If the target is within all positions of the conical scan of the beam a complete circle is shown on the screen, but if only a part of the scan covers the target a gap in the circle appears, the length of the gap being proportional to the error in the direction of the beam. The position of the gap indicates in which direction the beam is " off target ". A sense marker, in the form of a bright radial line on the display, is provided to ensure that the deflection coils producing the circular sweep on the cathode ray tubes are in angular alignment with the transmitted beam. The radius of the circle or arc is proportional to the slant range.

Provision is made for changing the range display according to whether the set is searching for or following a target, and a special feature is the use of an expanded time-base sweep to give the high incremental accuracy which is the primary requirement in wind-finding. For the search role, range is indicated on the tube in steps of 10 km., each represented by the maximum radius of the display. The steps are produced by a coarse time-delay which is introduced by the operation of a range increment switch. A range strobe marker in the form of a ring is provided on the screen and when this is set to coincide with the echo circle or arc the range is indicated on one of three meters in a range delay unit. This meter is graduated from 0 to 100 km. in 10 km. steps.

To follow a target after it has been located on the coarse-range display a variable fine time-delay, which is additive to the coarse delay, is switched into circuit. In the resulting expansion of the sweep the maximum radius on the screen represents 4 km. of range, centred at the range to which the strobe control has been set. The strobe marker ring remains stationary at half the maximum radius and the subsequent operation of the range control consists in bringing the inside edge of the target echo into coincidence with the marker. The range is then indicated by two meters in the fine time-delay circuit, one graduated from 1 to 14 km. and the other from 0 to 1 km. in 10 m. graduations. The range delay circuits are such that the incremental accuracy obtained from these two meters is independent of the absolute range accuracy of the coarse-range meter.

The range strobe is associated with the automatic gain control circuit in such a way as to make the latter effective only at ranges within $\pm$ 1 km. of the range of the strobe, so that unwanted signals and noise are reduced in proportion to the signal strength of the target echo.

4.5.2. Performance and accuracy

The maximum slant range specified for the equipment when tracking a corner reflector of 92 cm. hypotenuse length is 100 km. under normal propagation conditions. In specifying the accuracy of range measurement a distinction is drawn between the absolute measurement of slant range, which is required in the determination of height, and the measurement of the range interval between successive readings, which is more important for the wind determination. As stated above, in the ranging system employed the incremental accuracy is not limited by the absolute accuracy. For the latter the specification is that the error should not exceed 1 per cent. of the search range or 100 m., whichever is the greater, while for incremental accuracy it should not exceed 1 per cent. of the 10 km. increment. The accuracy of the azimuth and elevation measurements is specified by the requirement that the standard deviation of the errors should not exceed 0·13°.

Estimates of the errors of observation of a pre-production model of the W.F.1 radar have been derived from a series of twelve comparisons, made in 1957, with a Meteorological Office Type 1 radar. The results of the comparisons indicated that the specification is adequately met so far as the requirements of wind-finding accuracy are concerned. On the assumption that the observed errors were distributed equally between the two sets the standard deviations for errors in range is 24 m. and for errors in azimuth and elevation 0·13°. It is understood that an improved model of the Decca equipment is under development.*

4.6. NEW ZEALAND WIND-FINDING RADAR, TYPE M.E.7

4.6.1. Description

The M.E.7 equipment was developed as a wind-finding radar by the New Zealand Department of Scientific and Industrial Research from a 10 cm. radar set originally designed as a military height-finding radar. The main feature of the M.E.7 is the large parabolic aerial reflector which enables long ranges to be obtained with comparatively small targets without the need to use a high-power transmitter. A photograph of the aerial system is reproduced in Plate XIV.

The equipment is designed for permanent installation in a building, as shown in Plate XV, only the aerial system being outside. The paraboloid reflector is 4·6 m. (15 ft.) diameter but is truncated to 3 m. (10 ft.) in horizontal width. It is fed by a dipole at the focus and can be turned through 360° in azimuth and from 0° to 72° in elevation. Scanning is done by rotating the aerial through a small arc, first in azimuth and then in elevation, by manual operation of controls. The beam width (to half-power points) is about 1·5° in the vertical plane and 2·5° in the horizontal plane.

A rotating spark-gap modulator is used, giving pulses of 0·5 μ sec. width for use over short ranges and 3 μ sec. for use at long ranges, with a repetition frequency of 600 per sec. The transmitting valve is a magnetron giving a peak power of 120 kW. at a frequency of about 2800 Mc./s. The receiver incorporates a crystal mixer and local oscillator, and a two-stage I.F. preamplifier followed by a four-stage I.F. amplifier, the intermediate frequency being 24 Mc./s. with a band width of 2·5 Mc./s.

In the presentation unit, range is indicated on an A-type cathode ray tube display, by a time-base covering 0 to 120,000 yd. with a strobe selecting an expanded portion covering 10,000 yd. The strobe control potentiometer is calibrated at 5000 yd. intervals. Azimuth and elevation are shown on plan-position indicator

---

* After the foregoing description went to press the introduction of an improved model of the Decca wind-finding radar, known as Type W.F.2, was announced. This model differs from the Type W.F.1 mainly in the following respects. The peak power of the transmitter is 75 kW., instead of 25 kW., thus enabling slant ranges in excess of 150 km. to be achieved. The aerial is provided with a small auxiliary reflector which produces a broad beam (5°) that facilitates finding the target at the beginning of an ascent in adverse conditions. Improved instrumentation enables the equipment to be operated by one man, a means being provided for arresting the range, azimuth and elevation indicators when readings are to be taken. Lastly, a new design of cabin, with built-in air conditioning, improves the accessibility of the electronic units.

types of display in which the deflexion coils are rotated by servo-mechanisms from the aerial and are geared so as to rotate at twice or four times the speed of the aerial.

Operation of the equipment requires two observers, one of whom can undertake the plotting of the readings as well as reading the range tube.

Proposals for the design of two improved versions of the M.E.7 radar have been described by Hall and Barker[52].

4.6.2. Performance and accuracy

Information on the performance of the M.E.7 radar has been given by Ewing[53]. When the equipment is used with a corner reflector of 58 cm. hypotenuse length the maximum range from which useful signals can readily be obtained is about 110 km. At one of three New Zealand stations extreme ranges well in excess of this have been obtained. The accuracy of range measurement is approximately $\pm$ 90 m.

The accuracy that should be obtainable from the type of display provided for the azimuth and elevation indications is approximately 0·25°. Some comparisons with optical theodolite measurements showed that the errors of observation of the M.E.7 radar are such that the standard deviation is 0·5° in the case of azimuth and 0·3° for elevation.

4.7. RADAR REFLECTORS

4.7.1. General principles

The function of the balloon target or reflector in radar-wind measurements is to return echo signals back to the ground equipment from which the pulses are transmitted. The maximum range obtainable is limited by the intensity of the response, and the accuracy of the measurements depends to a large extent on the echo being steady. The chief requirements for a radar-wind target, therefore, are that it should be an efficient reflector or re-radiator of incident radiation at the wavelength of the transmitter and that its efficiency should not vary greatly with the angle of incidence. These two requirements are incompatible, since a target that reflects or scatters isotropically has a relatively low efficiency in the direction of the incident energy. In the design of a target, therefore, a compromise is necessary. It is, of course, essential that the target should be a good conductor of electricity. Other important considerations are that in order that the rate of ascent and the height performance of the balloon should not be much reduced the target should be as light as practicable and not add appreciably to the aerodynamic drag of the airborne assembly.

The efficiency of a non-absorbing object for returning radiation back in the direction of its source is expressed by its echoing cross-section, σ, which is defined as being equal to the cross-sectional area of a sphere, through its centre, which would give an echo of the same intensity. If the intensity of the radiation incident on the object is I_0 the intensity of the echo at a distance r is

$$I = \frac{I_0 \sigma}{4 \pi r^2}.$$

Unless the target is a sphere, which scatters the energy isotropically, the echoing cross-section will differ from the projected area S of the target perpendicular to the beam. The ratio σ/S can be regarded as a measure of the directional gain of a target.

Radar targets may be divided into two main classes, namely, resonators and reflectors. They will now be considered under these headings.

4.7.2. Resonators

The efficiency of a resonator as a target depends on the fact that, for its size, it absorbs and re-radiates a large amount of energy from an incident beam. Resonators used as radar targets usually take the form of half-wave dipoles. Such a dipole when orientated parallel to the electric field has an equivalent echoing area of $0{\cdot}86\lambda^2$ where λ is the wavelength. This must be multiplied by $\cos^2 \theta$ if the dipole is inclined at an angle θ to the direction of the field. If there are a large number, N, of dipoles with random orientations the average echoing cross-section of each will be one-third of that of a single one in the direction of the field. The effective echoing cross-section of the whole collection is therefore approximately $0{\cdot}29\, N\, \lambda^2$.

Dipole targets.—At wavelengths of 10 cm. and less the number of half-wave dipoles required to give adequate echoes at long ranges runs into hundreds. Their main use on these wavelengths is for screening military targets from radar beams. They are then known as " window " and are usually in the form of strips of metal foil or lengths of metallized thread, sometimes also known as " chaff ". Such dipoles have occasionally been used as targets in upper wind measurements on centimetric wavelengths by tracking clusters of them after ejection from a balloon, aircraft or rocket.

The use of dipole targets for routine wind measurements is generally restricted to radar systems operating on wavelengths exceeding 1 m. The most usual arrangement consists of three conductors, each a half wavelength long, fixed mutually at right-angles. In an alternative arrangement the three conductors are disposed symmetrically in a horizontal plane. The conductors may be made of aluminium foil or wire supported on light strips of wood or similar material.

4.7.3. Passive reflectors

Targets which depend on reflection or scattering rather than absorption and resonance for their efficiency must have dimensions large compared with the wavelength, but no advantage is gained by increasing the size so much that the target extends beyond the beam width. The echoing cross-sections of the simpler forms of targets can be calculated, the simplest case being that of a conducting sphere, which has the property of scattering equally in all directions. Its echoing cross-section is independent of the wavelength, being the same as its cross-sectional area, namely, πa^2, where the radius a is large compared with the wavelength. The gain, therefore, is unity and the sphere is not a very efficient target. It has been used, however, as a standard in the measurement of the power of a radar set. Ellipsoidal targets, the principal radii of curvature, a_1 and a_2, of which do not differ greatly, have echoing areas given by $\pi a_1 a_2$ which are again independent of the wavelength. On the other hand the wavelength does affect the performance of a cylindrical target. For a cylinder of length l and radius a, with its axis normal to the beam, the echoing area is $2\pi a l^2/\lambda$.

Flat plates form very efficient reflecting targets provided they are fixed in a position normal to the direction of the incident radiation. The echoing cross-section of a flat plate of area S is given by $4\pi S^2/\lambda^2$ but the large echo thus produced is restricted to a very small range of angles about the normal. If, however, the radiation is reflected in turn from three plates mutually at right-angles the emergent ray returns in the same direction as the incident ray, as illustrated in Fig. 12. This arrangement produces a large echo over a considerable range of angles of incidence and is the principle of the corner reflector, an everyday application of which is to be seen in the rear reflectors on road vehicles. The echoing cross-section of a corner reflector is the same as that for a flat plate normal to the beam but S is now the effective aperture of the corner reflector over which rays are reflected. For triple reflection this is a maximum when the direction of propagation makes equal angles with all three surfaces; this angle is 35°. If the corner is formed by three right-angled triangles joined along their short sides the effective maximum aperture S is a regular hexagon contained within, and two-thirds the area of, the equilateral triangle formed by the hypotenuses of the right-angled triangles. If l is the hypotenuse length, the area S of the hexagon is $l^2/2\sqrt{3}$ and the maximum echoing cross-section is, therefore, $\pi l^4/3\lambda^2$. As the angle of incidence of the beam departs from the axis of symmetry the amount of energy reflected diminishes; it falls to about half the maximum at an angle of 20° from the axis. To ensure the return of strong echoes from all directions the reflectors used in wind-finding generally take the form of an assembly of four corners with a common apex and sometimes eight corners are used. The conducting surfaces may be made of metal foil (usually aluminium) on a paper base or of wire or metallized fabric netting. When net is used the size of the mesh should not exceed one-twentieth of the wavelength for which the reflector is to be used. A disadvantage of metal foil is that it causes an appreciable aerodynamic drag, but this can be reduced by perforating the foil with some large holes. The material is mounted in a light wooden or metal framework, designed to be readily assembled into the required shape, with the planes mutually perpendicular.

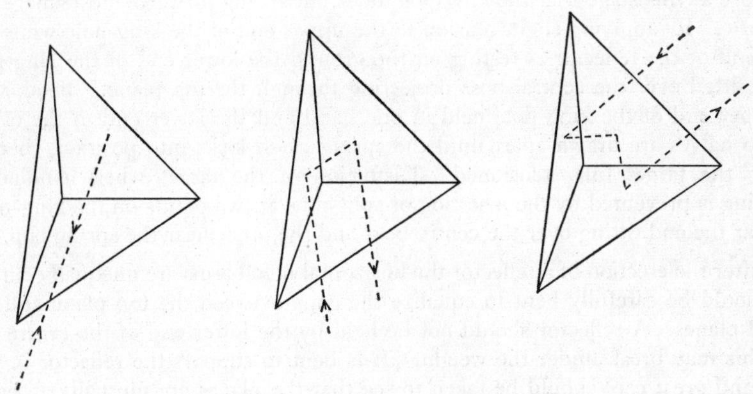

FIG. 12—PRINCIPLE OF THE CORNER REFLECTOR

Meteorological Office radar reflectors.—These consist of three planes one of which is a square that is intended to remain nearly horizontal in flight. Underneath the square, right-angled triangles are fitted along the diagonals so as to form four corners. When the top plane is horizontal, the efficiency of the reflector is at a maximum for beams with an elevation of 35°. The conducting planes are made of metallized net with a mesh of not less than two per centimetre (for a wavelength of 10 cm.) and are attached to a light framework in such a way that they can be folded or rolled, for storage, into a small bulk. When assembled for use the planes should be mutually perpendicular to within 1° and their surfaces should be plane to within 6 mm. The reflector is made in two sizes. In the larger size, known as Mark 3 (Stores Ref., Met. 10090), which is illustrated in Plate XVI, the square top plane has a side 1·37 m. long. This reflector weighs about 600 gm. and its performance is such that with a Type 1 radar it will give an average signal-to-noise ratio of 2 : 1 at a slant range of 40 km. The smaller reflector, known as Mark 5 (Stores Ref., Met. 10492), has a top plane with sides 0·69 m. long and it is only used for soundings to relatively short ranges.

The procedure for erecting the reflectors differs according to the design of the framework adopted by different makers. In one make, identified by the letter A, the triangular net planes are first unwrapped from around the centre pole, taking care that the stay wires are in the grooves at the top of the pole and not wound round the ends of the arms. The upper boss is then moved about three-quarters of the way up the centre pole and the stay wires again examined to ensure that they do not foul the arms. The lower boss is released by depressing its clip and the upper boss is pushed home until it engages the upper clip. Then the lower boss is pulled down until it engages at the required tension on one of the three steps of the lower clip. The reflector is then ready for use.

In erecting a reflector with identification letter B, care must be taken, when unwrapping the triangular planes from around the centre pole, that the loose king-pole does not tear the net and that the four wires attached to it are not entangled. The target is then inverted by holding it in one hand by the boss projecting from the centre of the top plane, the loose king-pole being held in the other hand, taking great care at this stage that the king-pole does not spring forward and damage the net fabric. By applying slight tension to the upper end of the king-pole while the lower end of the reflector is resting on the ground the lower end of the king-pole can be fitted over the central boss projecting through the top plane. Then, with the upper end of the king-pole held in one hand and the lower end in the other, the two halves are drawn apart until the spring catch locks into position, thereby making the target fully tensioned. Distortion of the target when handled at launching is prevented by the insertion of split-pins at two points on the king-pole, one near the end fitting over the centre boss and the other near the spring clip.

If after the erection of a reflector the horizontal metal arms are unequally curved they should be carefully bent to equalize the gaps between the top plane and the vertical planes. A reflector should not be held by the lower end of the centre rod since this may break under the weight. It is bent to support the reflector in two places and great care should be taken to see that the planes are mutually at right-angles and flat to a high degree of accuracy.

Decca corner-reflector system.—This system, which is designed for use with 3 cm. radar, employs two standard sizes of triangular units of pre-stressed paper-backed metal foil on a light-weight metal wire frame. The units can be clipped together, with simple clips applied by a hand tool, to give two different types of corner clusters in two sizes. For the larger sized four-cornered type, the top plane is a square of side 92 cm., the other two planes being right-angled triangles, the hypotenuses of which fit along the diagonals of the top plane. In the smaller four-cornered reflector the top plane has sides 65 cm. in length. These reflectors can be converted to eight-cornered types by adding another cluster of four corners to the upper side of the top horizontal plane. Some advantage may be gained by the use of eight corners when observations are being made at long ranges and low elevations, since the swinging of the target may reduce appreciably the efficiency of the lower half of the target. The theoretical maximum echoing cross-sections of the large and small targets are 725 m.^{2} and 175 m.^{2} respectively.

Expanding balloon covers.—Attempts have been made to use a balloon as a spherical radar target, for example, by impregnating the rubber with conducting material. It was found that as the balloon expanded the conductivity became too low, owing to the separation of the conducting particles, to give satisfactory echoes. Useful results have been obtained, however, by covering the balloon with a shroud of flexible metallized net. This device, which was originally developed by the Admiralty, has the advantage of adding very little to the aerodynamic drag of the balloon and of being much more easily handled than large corner reflectors. It was developed for use with 100 gm. balloons in soundings up to about 15 km. The use of such covers for larger balloons would be uneconomical owing to the large amount of material required.

The cover, which weighs 70 gm., is designed to act, when fully stretched, as a hemispherical reflector of 1·5 m. diameter, but an elastic cord through the hem secures it on to the balloon at smaller diameters. As the balloon expands from its initial inflated diameter of about 1 m., the hem slips smoothly over the surface, and after the hem is fully stretched it continues to hold the cover like a cap on the balloon, which goes on expanding through the open aperture of the cover.

In the initial stages of an ascent the cover acts more as a convex reflecting surface, but after some expansion it can act as a concave reflector, with considerably higher efficiency than that of a sphere of comparable size. In order to ensure that the cover presents its inner surface to the radar beam for the greater proportion of the period of the balloon's oscillations a stabilizing weight of about 80 gm. is suspended from the balloon on a 10 to 20 m. string.

4.7.4. Summary of radar target characteristics

The theoretical formulae for the echoing cross-sections of the various types of target are summarized in Table VII.

TABLE VII—ECHOING CROSS-SECTION FORMULAE

Target	Size	Orientation	Projected area	Echoing cross-section
Dipole ..	Length $\lambda/2$	Parallel to field	...	$0{\cdot}86\lambda^2$
N dipoles ..	...	Random	...	$0{\cdot}29N\lambda^2$
Sphere ..	Radius a	...	πa^2	πa^2
Flat plate ..	Area S	Normal to beam	S	$4\pi S^2/\lambda^2$
Corner reflector	Hypotenuse l	Corner symmetrical to beam	$l^2/2\sqrt{3}$	$\pi l^4/3\lambda^2$

Some numerical data are given in Table VIII. In addition to the theoretical values of the echoing cross-sections, values of maximum ranges relative to those of spheres of diameter ten times the wavelength are given. The formula for echo intensity given in Section 4.1.2 shows that the maximum range is proportional to the fourth root of the echoing cross-section. It should be emphasized that the relative values for the 10 cm. wavelength are not comparable for 3 cm. since radar characteristics may be very different for different wavelengths. It will be noticed that the maximum range for a perfectly accurate corner reflector is proportional to its linear dimensions.

TABLE VIII—ECHOING CROSS-SECTIONS AND RELATIVE MAXIMUM RANGES

Wavelength	Reflector	Size	Echoing cross-section	Relative maximum range*
cm.			m.²	
100	Dipole	50 cm. long	0·86	...
10	Sphere	100 cm. diameter	0·79	1·0
10	Flat plate (normal to beam)	100 cm.² area	1250	6·3
10	Corner, Mark 3	137 cm. hypotenuse	370	4·7
10	Corner, Mark 5	69 cm. hypotenuse	23	2·3
10	Balloon cover	150 cm. diameter	...	(2·8)
3	Sphere	30 cm. diameter	0·071	1·0
3	Corner, Decca 1	92 cm. hypotenuse	725	10·2
3	Corner, Decca 2	65 cm. hypotenuse	175	7·1

* Values for 10 cm. are not comparable with those for 3 cm.

4.8. RADAR-WIND COMPUTING DEVICES

The methods and equipment used for computing upper winds from pilot-balloon ascents, described in Section 3.5, are generally applicable to computations from radar-wind observations. For example, a pilot-balloon slide-rule may be used for evaluating the wind speed and direction and the height of the balloon from the radar observations of slant range, azimuth and elevation. A graphical method is to be preferred, however, since it facilitates the process of smoothing which is desirable in order that the effects of random errors of observation and of atmospheric turbulence may be eliminated or reduced. Moreover a graphical method allows rapid selection of the points of significant wind change which, together with the winds at standard heights, are reported. In Meteorological Office practice a combination of slide-rule computation and graphing technique is employed, the slide-rule being used for evaluating the horizontal range and height and the graphical process for plotting the plan-position of the balloon.

4.8.1. Meteorological Office radar-wind slide-rule, Mark 1

This slide-rule (Stores Ref., Met. 10496) has been specially designed for evaluating $R \cos E$ and $R \sin E$ from the radar readings of slant range R and elevation E, thus giving the horizontal range and the height. It is illustrated in Plate XVII and consists of a stock graduated with logarithmic scales of sines and cosines, and a slider carrying logarithmic scales of slant range (metres), each scale being 21·5 in. long. The logarithmic cosine scale is on the upper part of the stock and covers, from left to right, angles from 84·3° (approximately) to 0°. Index arrows are placed above the graduations at each end, these graduations corresponding to cosines having the values 0·1 and 1·0. The lower part of the stock carries the logarithmic sine scale running from 35° to 90° on the left and from 5·7° (approximately) to 35° on the right. The common index graduation at 90° corresponds to sines having the values 1·0 and 0·1 and lies immediately opposite the 80° graduation on the cosine scale. On the upper edge of the slider is a simple logarithmic scale covering slightly more than the single decade 1 to 10 and graduated accordingly. The length of this scale between the graduations 1 and 10 is exactly the same as that between the index marks on the stock corresponding to the values of 0·1 and 1·0 for the cosines.

A logarithmic scale of the same length on the lower edge of the slider covers slightly more than the single decade from 0·574 to 5·74. Two cursors are provided, one for the upper pair of scales and the other for the lower pair.

The first operation in using the rule is to set the slant range on the upper scale B of the slider against the right-hand index mark of the cosine scale A. Then, with the upper cursor set at the angle of elevation on the scale A the horizontal range is indicated on the slider scale B, and with the lower cursor set to the angle of elevation on the sine scale D the height is indicated on the lower scale C of the slider. If no part of scale C falls opposite to the angle of elevation on scale D the first operation must be repeated using the left-hand index mark of scale A. It should be noted that the latter does not correspond exactly to the angle 84·3° and neither does the mark nor scale D opposite to 80° on scale A correspond exactly to the angle 5·7°. The range of angles covered by the rule is adequate for all practical purposes.

In a Mark 2 version at present under consideration it is proposed to number the cosine scale additionally in terms of sines and the sine scale additionally in terms of cosines, in order to obviate the need for repeating the settings that would otherwise be necessary with certain combinations of range and elevation.

Heights $R \sin E$ must be corrected for the curvature of the earth's surface. The correction is $+ (R^2 \cos^2 E)/2r$, where r is the radius of curvature of the surface, i.e. 6380 km. approximately. The effect of atmospheric refraction is to make the angle of elevation, and therefore the height, too great. An idea of the errors due to a standard atmosphere has been given in Section 4.1.2. The variations in error due to departures of the atmosphere from standard will, of course, be smaller. For wind-finding with the equipment at present in use no correction for the effect is applied, but if heights and ranges attained are substantially increased allowance for refraction will become necessary.

4.8.2. Meteorological Office radar-wind plotting tables

The Mark 1 design of plotting table (Stores Ref., Met. 269) consists of a wooden table top, 1·27 m. wide and 2·18 m. long, to which is permanently attached a ruled paper chart, the table being supported by a galvanized mild-steel framework. The chart is ruled in polar co-ordinates, with the origin midway between the longer sides and 30 cm. from one of the shorter sides. Formerly, circles or arcs of circles were drawn, with the origin as centre, having radii increasing in steps of 18·3 cm., each step representing steps of 10,000 yd. in range, but these are no longer required. Radial lines are provided at 1° intervals except near the origin, where they are drawn at 5° or 10° intervals. When in use the chart is overlaid with tracing paper on which the plan-position of the balloon is plotted. Supports and clamps for the tracing paper are fitted so as to provide a ready means of fixing the paper in position and of adjusting its position between ascents.

Three accessories are available for use with plotting tables. One is a plotting rule (Mark 1, Stores Ref., Met. 10512) for marking off horizontal ranges from the origin. This rule is made in sections which can be rigidly fitted together to the required length. The sections are made of perspex held in a metal frame and a pivot point is provided for accurately locating one end of the rule at the origin. A scale of kilometres, divided into 50 m. intervals and reduced in the ratio 1 : 50,000, is engraved along one edge of the perspex. There is also a line drawn along the length of the rule making an angle of 1° with the graduated edge. At three places along the length this angle is sub-divided into 0·1° intervals. The second accessory is a scale for measuring wind speed from the plotted chart. It is known as the radar-wind scale Mark 1 (Stores Ref., Met. 2663) and is merely a transparent rule marked off in knots to correspond with a scale of 1 in 50,000, and a 3 min. interval of measuring time. The third accessory is an Admiralty pattern drafting mechanism (Mark 4) to which is fitted the radar-wind scale Mark 1. The mechanism allows the wind scale to be manoeuvred over the wind table, carrying with it a scale of azimuth locked to the orientation adopted for the table. The wind rule is aligned with the balloon track to obtain the wind speed, as already described, and reference to the azimuth scale gives immediately the corresponding wind direction.

In the most recent design of plotting table (Mark 2, Stores Ref., Met. 10568) the chart is ruled on the underside of thin transparent plastic material and is more durable and less liable to shrinkage than a paper chart permanently attached to wood. The plastic chart is secured to the table top only at the origin but it is prevented from moving by aluminium beading at the edges of the table. This method of fitting it facilitates the renewal of the chart. The rulings on the chart are confined to the radial lines, the circles being omitted. A container for holding a roll of tracing paper is fitted at one end of the table, thus providing a convenient means of dispensing the paper for plotting.

For radar-wind ascents from Meteorological Office ocean weather ships the plotting is done on a plain paper chart (45 × 31 in. in size) with the aid of an Admiralty pattern drafting mechanism (Mark 4), allowance for the ship's movement being made by successive displacement, minute by minute, of the origin from which plan ranges are set off. An alternative method of plotting which is advantageous in bad weather is the use of a shore-station plotting table, when allowance for ships movement is made either by setting off from the plotted balloon plan-positions or by solution of the appropriate vector triangles.

CHAPTER 5

SECONDARY RADAR FOR UPPER WIND MEASUREMENT

5.1. THEORETICAL CONSIDERATIONS

It has already been explained that secondary radar differs from primary radar in using a transponder to regenerate the signal received by the airborne equipment from the ground transmitter. The chief advantage of secondary radar is its capability of operating over long ranges with a much lower expenditure of power (though not necessarily at lower cost) than is practicable with primary radar. A further attractive feature is the possibility of designing a transponder to act also as a telemetering system for other information in addition to the data necessary for determining its location.

It does not require much power for a transponder to produce a greater response than an echo from an efficient reflector at the same range. For example, the maximum power echoed by a 1·37 m. corner reflector (echoing cross-section = 370 m.^{2}), receiving radiation from a Meteorological Office Type 1 radar (peak power $P = 350$ kW. and aerial gain G about 1000) at a range R of 80 km., is given by $P\,G\,\sigma/4\,\pi\,R^2$ and amounts to about 1·5 W. A transponder power well in excess of this is quite practicable and therefore much longer ranges should be attainable if sufficient gain can be provided.

It is not necessary for the outgoing and incoming links in a secondary radar system to operate on the same frequency, and as a general rule different frequencies are used. The theoretical power requirements of the two links may be determined by the following considerations. If P_{tg} and G_{tg} are the power and gain of the ground transmitter the intensity of the radiation at a range R will be

$$I_a = \frac{P_{tg}\,G_{tg}}{4\,\pi\,R^2}\,.$$

The power received by the transponder at this range is $I_a\,\lambda_t^2\,G_{ra}/4\,\pi$, i.e. $P_{tg}\,G_{tg}\,G_{ra}\,\lambda_t^2/16\,\pi^2\,R^2$, where G_{ra} is the gain of the transponder receiving aerial and λ_t the wavelength of the ground transmitter. The maximum range of operation of the ground-to-air link (in the absence of noise) is therefore reached when this power falls to the minimum power P_{0a} required to trigger the transponder. Similarly, the maximum range of operation of the air-to-ground link is reached when the power received at the ground receiver, viz. $P_{ta}\,G_{ta}\,G_{rg}\,\lambda_a^2/16\,\pi^2\,R^2$, is the minimum P_{0g} required to produce a signal. In this case P_{ta}, G_{ta} and λ_a are the power, gain and wavelength of the transponder transmitter and G_{rg} is the gain of the ground receiver aerial. If the two links are designed to give the same maximum range, the determining factors are theoretically related by:

$$\frac{P_{tg}\,G_{tg}\,G_{ra}\,\lambda_t^2}{P_{ta}\,G_{ta}\,G_{rg}\,\lambda_a^2} = \frac{P_{0a}}{P_{0g}}\,.$$

In practice, however, other factors may influence the maximum ranges.

The transponder consists essentially of a receiver, a triggering circuit and a transmitter, together with appropriate aerials. There is always a small delay, usually of a fraction of a microsecond, between the receipt of the incoming signal and the emission of the outgoing reply. For the purposes of accurate wind measurement it is important that this delay should remain constant as well as small.

The earlier applications of secondary radar to wind-finding, such as the German " Fledermaus " system and the Swiss " Telesonde ", used continuous wave, and not pulse, transmission. They were designed for the measurement of slant range and azimuth but not elevation, this being obtained from pressure–height measurements of a radio-sonde. The secondary radar wind system developed for the Meteorological Office uses pulse transmission and is designed for the measurement of elevation as well as azimuth and range.

5.2. THE METEOROLOGICAL OFFICE RADAR-THEODOLITE SYSTEM

5.2.1. General description

The basic design of this system, which was undertaken in the Ministry of Supply to meet the requirements of the Meteorological Office, was described by Jones, Hooper and Alder[54]. Its development for the Meteorological Office was undertaken by the Mullard Research Laboratory and Goddard and Dell[55] have described the prototype equipment. The system was designed for two main functions: (*a*) to measure the wind by the secondary radar principle and (*b*) to measure pressure, temperature and humidity and transmit these measurements to the ground equipment through the transponder channel. The first function may be used alone, while the second one was intended to be used by the addition of a sonde unit to the transponder circuit. A feature of the equipment is that it is completely automatic in operation. As we are only concerned here with the use of the system for wind measurement, further discussion of the sonde unit is deferred until Chapter 7.

The radar-theodolite system was required to track satisfactorily a balloon-borne transponder to distances up to 185 km. (100 nautical miles) and thus enable the wind speed and direction to be determined, from the surface to approximately 25 km. (30 mb.), with a specified accuracy. It should be mentioned that more recently the essential height requirement has been increased to 30 km. and extension to greater heights would be desirable. The equipment, which is designed for operation in a permanent building, is illustrated in Plates XVIII and XIX and Fig. 13.

The ground equipment comprises the aerial unit, shown in Plate XVIII, radar transmitter and receiver, in Plate XIX, and computing and recording equipment. A pair of Yagi aerials transmits interrogating pulses of 2 μ sec. width on a carrier frequency of 152·5 Mc./s. (2 m. wavelength) with a peak power of 50 kW. and a pulse recurrence frequency of 404 per second. This aerial array is fitted to the same mount as the receiver aerial, in order to direct the maximum energy towards the airborne unit, but it does not play any essential part in direction-finding. The signals received at the ground from the transponder are on a frequency of 2850 Mc./s. (10·5 cm. wavelength) and are received on a nutating dipole aerial in a paraboloid reflector. The conical scan so produced generates an error signal when the aerial is " off target " and this signal is fed into a servo-system which automatically aligns the aerial in the direction of the airborne transponder.

Azimuth and elevation angles from the aerial unit, together with slant range from the transit time of the pulse to and from the transponder, are fed to a computer and also to a display unit where slant range is displayed on a cathode ray

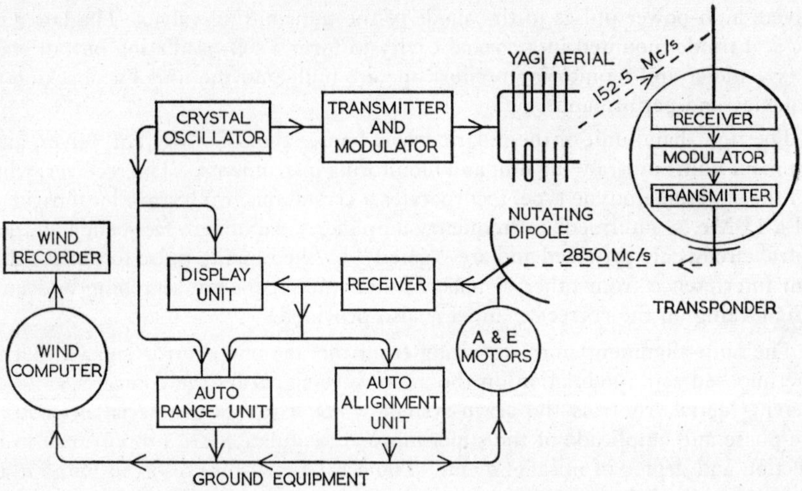

FIG. 13—BLOCK DIAGRAM ILLUSTRATING THE RADAR-THEODOLITE SYSTEM

tube and azimuth and elevation on fine and coarse magslip indicators. The computer and recorder units calculate and record the wind speed and direction and the height of the airborne unit at a given time. The transmitter, receiver, display unit and power supply control panel are mounted in one console (Plate XIX) and the computer and recorders form separate units. The transponder (Plate XIX) consists essentially of a receiver for the 152·5 Mc./s. signals, a modulator and a 2850 Mc./s. transmitter. The aerials are in the form of quarter-wave rods projecting upwards and downwards from the thermally insulating case in which the transponder is housed. Power is supplied from a 7 V., 2·5 amp. battery with a vibrator-transformer for producing high-tension voltages. It should be mentioned here that in the early stages of the development a continuous wave method was tried but since it produced serious frequency modulation of the centimetric transmitter, it was decided to use pulse technique.

5.2.2. Ground equipment—technical details (see Fig. 13)

Aerial unit.—The pair of Yagi transmitting aerials, mounted on either side of the receiver paraboloid and connected in parallel, have a power gain of about ten. They have a beam width (at half-power) of 25·5° in the horizontal and 50° in the vertical planes. The receiving aerial has a 5 ft. diameter paraboloid reflector and a beam width of about 5°; it is similar to that used in the Meteorological Office Type 3 radar (Section 4.3). The dipole nutates at approximately 11 revolutions per second and the half-angle of the conical scan is 2°. Nutation rather than rotation of the dipole is required because the polarization of the transponder signal is substantially vertical.

Radar console.—The left-hand part of the console houses the 152·5 Mc./s. transmitter. This produces a peak power of 50 kW. in pulses of 2 μ sec. duration. A crystal-controlled oscillator provides the pulses for triggering the transmitter (and other parts of the equipment) at a repetition frequency of 404 per second and it also provides calibration markers for the range display tube. In the transmitter a blocking oscillator is triggered by the pulses and in turn generates a pulse with the required wave form. This, when amplified, drives a modulator valve which delivers high-power pulses to the anode of the transmitting valve. The latter is a disc-seal triode mounted in a coaxial cavity to form a self-oscillating output stage. A wave-meter and monitoring oscilloscope are built into the unit for checking the frequency and performance.

The right-hand unit of the radar console houses the 2850 Mc./s. receiver, auto-alignment unit, auto-ranging unit and monitoring instruments. The receiver, which is of the superheterodyne type, incorporates a crystal mixer, klystron local oscillator and a 13 Mc./s. intermediate-frequency amplifier. Automatic frequency and gain control circuits are provided and are " gated " by the ranging pulse to protect them from interference from other signals. Automatic frequency searching and automatic locking on the correct channel is also provided.

The auto-alignment unit, operating from the alignment error signals that are superimposed as a modulation on the microwave signal by the conical scan of the receiving aerial, controls the servo-system which corrects the aerial orientation. The phase and amplitude of the superimposed modulation are proportional to the direction and degree of misalignment. The error signals are resolved into left and right and up and down components and after amplification are fed into a closed loop servo-system which includes the aerial driving motors and tachometer generators. The latter produce voltages proportional to the rates of change of azimuth and elevation for use in the wind computer and also in the feed-back circuit of the servo-system. The aerial driving motors operate over a speed range of 1 to 10,000 r.p.m. and the sensitivity of the system is such that a misalignment of less than one minute of arc is sufficient to produce full torque in the driving motors.

The auto-ranging unit generates a strobe pulse, the time delay of which can be varied manually or automatically, and the strobe can thus be kept superimposed on the transponder signal as the range increases. The timing of the delay between the transmitter pulse and the strobe pulse is measured by comparison with a calibration signal of frequency 808 kc./s. produced by the crystal oscillator. The time interval between consecutive calibration pulses corresponds, therefore, to 185 m. (0·1 n. mile) in range, and they are locked in phase with the pulse modulation of the ground transmitter since this is triggered, with a repetition frequency of 404 pulses per second, by the same crystal oscillator. Precise adjustment of the strobe pulse within the 185 m. interval to which it is locked is effected automatically by a motor-driven phase-shifting transformer which advances or retards the phase of the calibration pulse train until it coincides with the range of the transformer signal. One complete rotation of the phase-shifting transformer introduces a phase change of 2π, corresponding to a range interval of 185 m. A tacho-generator coupled to the rotor of the transformer provides a voltage proportional to the rate of change of range, i.e. to the radial velocity of the airborne unit. This voltage, together with a voltage controlled by the strobe delay and therefore proportional to the range, is fed to the wind computer.

The middle section of the radar console contains the power supply control panel and the display unit. The latter provides two cathode ray A-scan traces (corresponding to coarse and fine range scales), on which are displayed the signal pulse and the ranging pulse generated in the auto-ranging unit. The traces have range markers superimposed on them at appropriate intervals; in the case of the fast trace the smallest intervals correspond to 185 m. (0·1 n. mile). Four magslip repeaters indicating the azimuth and elevation of the aerial orientation to the nearest 0·1° are incorporated in the display unit and manual controls are provided to allow the aerial to be " put on " to the airborne unit at the beginning of an ascent.

The wind computer and recorder.—The basis of the method for the automatic computation of the wind is the formulation of the tangential and radial components of the wind and the solution of the vector triangle. These components are expressed by:

$$V_T = R\frac{dA}{dt}\cos E$$

$$V_R = \frac{dR}{dt}\cos E - R\frac{dE}{dt}\sin E,$$

where R, A and E are the slant range, azimuth and elevation, which, together with their rates of change, comprise the basic data provided by the auto-ranging and auto-alignment units. The height of the airborne unit is obtained from the expression:

$$H = R\sin E + \frac{R^2\cos^2 E}{D},$$

the second term being an approximate expression for the correction due to the curvature of the earth, the diameter of which is D.

The computer is of the analogue-type employing servo-controlled precision potentiometers for multiplication and magslip resolvers for solving the vector triangle. It should be noted that the wind direction is computed relative to the azimuth of the airborne unit and is, therefore, corrected by means of an azimuth servo to give a direction relative to true north. Wind speed, direction and height of the balloon are recorded on a conventional pen recorder of the strip-chart type.

5.2.3. The transponder

The transponder is housed in a thermally insulating case made of expanded ebonite about 2·5 cm. thick. This is suspended from the balloon by a quarter-wave aerial on which the 152·5 Mc./s. interrogating pulse is received. The pulse is fed into a receiver which has a radio-frequency amplifier and a detector; sub-miniature valves are used. The output from the receiver triggers a blocking-oscillator which generates a 1 μ sec. pulse of 50 V. amplitude and modulates the 2850 Mc./s. transmitter. Triggering sensitivity is about 3 mV. at the receiver terminals and the band width is about 4 Mc./s.

The transmitter is a disc-seal triode valve, of type CV273, in a coaxial cavity. With 800 V. on the anode and pulses of 50 V. from the blocking-oscillator on the grid a peak power of at least 30 W. is obtained. The output is fed into a quarter wavelength aerial mounted vertically at the base of the transponder and projecting through the cover. The aerial is provided with a counterpoise and radiates in a generally downward direction, with nominally vertical polarization when the unit is not swinging on its suspension.

Power supplies for the transmitter and receiver are provided by primary lead–acid cells. High tension supply of 100 V. for the receiver and modulator and 800 V. for the transmitter are derived from a vibrator-rectifier unit.

5.2.4. Performance and accuracy

Information on the performance and accuracy of the radar theodolite system is based on trials of the prototype equipment installed at the Meteorological Office upper air station at Crawley. After these trials some modifications to the design were proposed. It is emphasized, therefore, that the performance of the prototype would be improved upon in an equipment incorporating the modifications.

The specification required the equipment to be capable of operating over all distances up to 185 km. and of determining the wind speed and direction at all heights from the surface to the level of 30 mb. (approximately 25 km.) with an accuracy such that at least on 90 per cent. of all occasions the vector error does not exceed 2·7 kt. (5 km./hr.). This is a very high accuracy to attain at long ranges.

The trials included measurements with transponders released from a site 60 km. distant from the radar-theodolite and tracked simultaneously with a Meteorological Office Type 1 radar at the release point. The results, which referred to dial readings of the radar theodolite display, were mostly confined to ranges between 20 and 75 km. They showed the total error in range measurement to be much less than 100 m. and the standard deviation of range measurement to be less than 35 m. There was a mean error in azimuth of about + 0·2° but the greater part of this was due to error in the dial setting, which could be eliminated by adjustment. The standard deviation of the azimuth measurements was about 0·05°. A mean error in elevation of about + 0·08° was found to be due to misalignment of the electrical and optical axes of the aerial system. The standard deviation of elevation measurements is about 0·03°. If these estimates of standard deviation are applied in formula (i) of Section 2.3.2 to the case in which the transponder ascends at 6 m./sec. in a mean wind of 36 kt. the standard vector error of wind measurement at a height of 20 km. (and therefore a range of 60 km.) is calculated to be 2·8 kt. This error, however, refers to winds computed from dial readings which do not make use of the rate of change information that is provided for the automatic computer. The latter therefore should give higher accuracy. Comparisons of automatically computed winds with those computed from half-minute readings of the dials showed mean differences for an ascent with a mean speed of 28 kt. to be — 1 kt. in speed and + 2° in direction. It is probable that the system comes very near to meeting the specification of accuracy of wind measurements.

Precise information on the maximum range of the system is not available. In the trials, marked fading occurred on some occasions at comparatively short ranges but on other occasions satisfactory records were obtained at ranges near the specified figure of 185 km. Satisfactory performance of an automatic computing and recording equipment is dependent on the received signal being reasonably steady and a signal-to-noise voltage ratio of at least 5 : 1 being available throughout the whole range to be covered. Although the power transmitted in the outgoing link is quite adequate to trigger the transponder under normal conditions, failure of this link is a probable cause of inefficient operation at high angles of elevation, when the airborne unit with a nearly vertical receiving aerial is being interrogated by a nearly horizontal transmitting aerial.

It may be of interest here to give theoretical estimates of the mean signal levels to be expected in the two links. Taking first the outgoing 2 m. wavelength link, with a peak power P of 50 kW., transmitter aerial gain G_t of 10 and airborne receiver aerial gain G_r of 1, the power received by the transponder at a range R of 185 km. is given by:

$$P_r = \frac{P G_t G_r \lambda^2}{16 \pi^2 R^2} = 3 \cdot 6 \times 10^{-7} \text{ watts}$$

for a signal propagated horizontally (normal to the receiver aerial). The triggering sensitivity of the transponder is about 2·5 mV. at the terminals of the quarter-wave aerial, the impedance of which is about 40 Ω, and so the minimum power to trigger

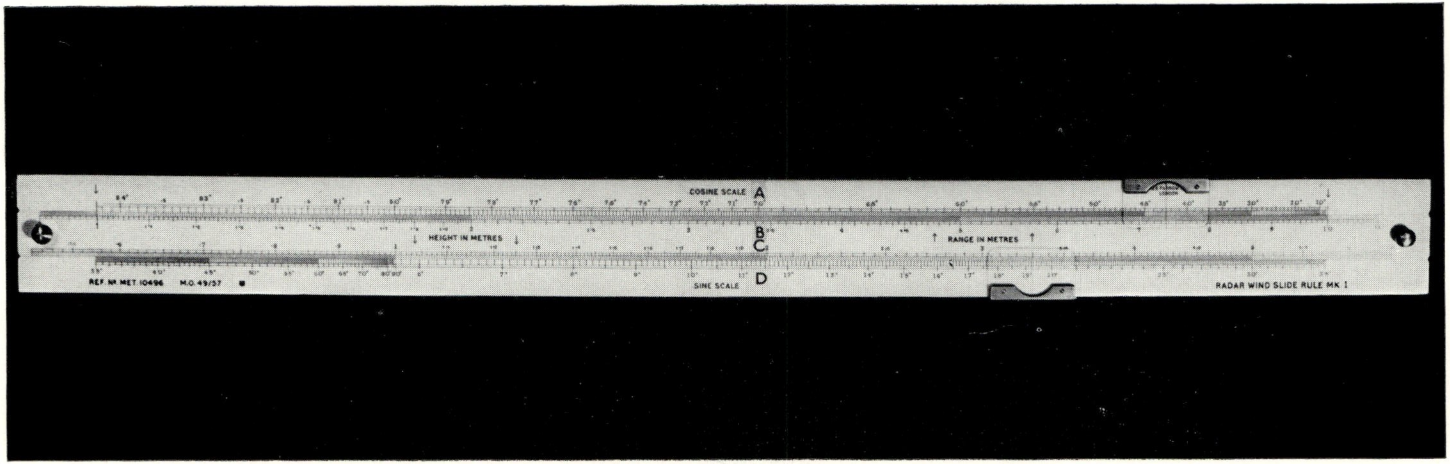

METEOROLOGICAL OFFICE RADAR-WIND SLIDE-RULE, MARK 1

Plate XVIII

By courtesy of Mullard Ltd.

RADAR-THEODOLITE AERIAL UNIT

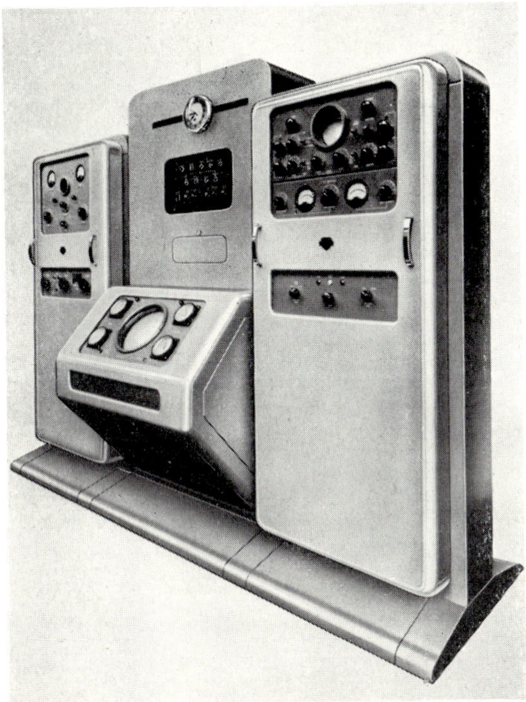

By courtesy of Mullard Ltd.

RADAR-THEODOLITE CONSOLE

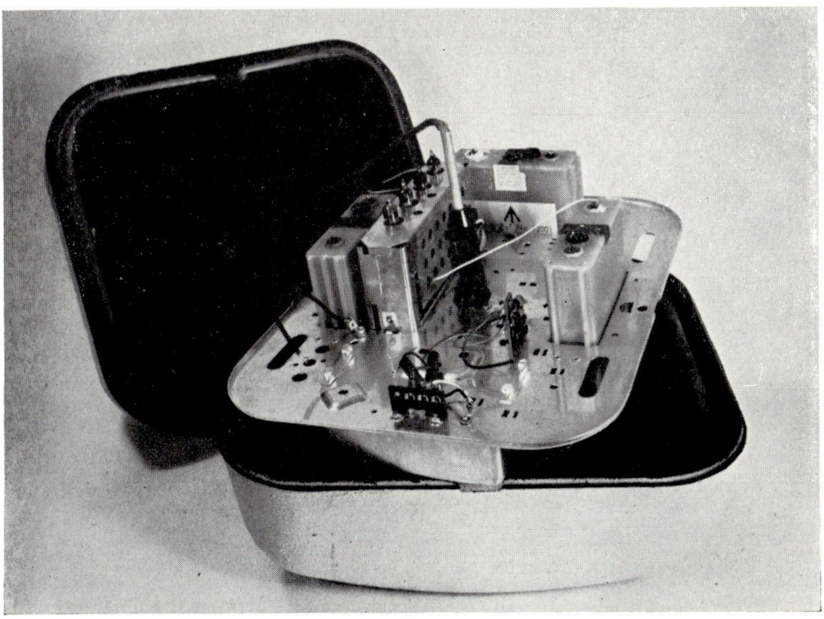

TRANSPONDER USED IN THE RADAR-THEODOLITE SYSTEM

Plate XX

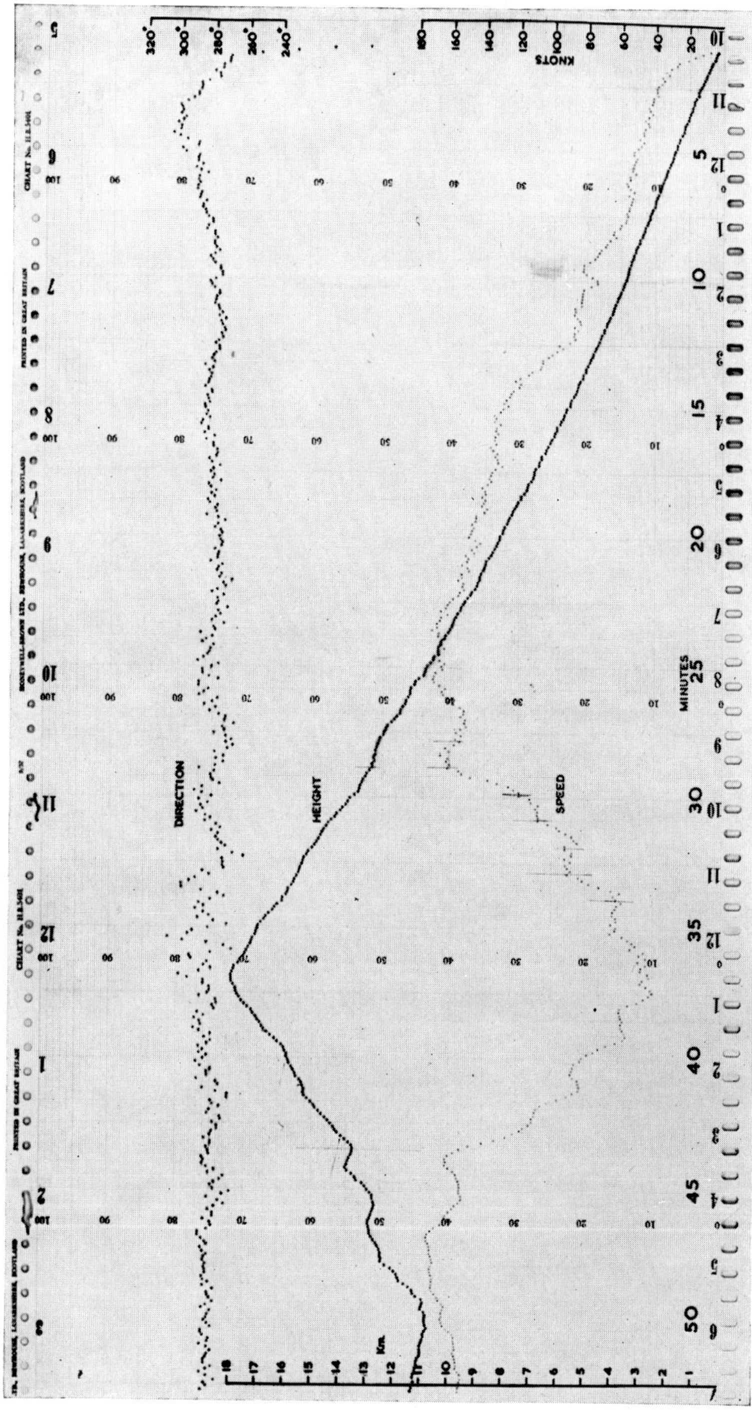

RADAR-THEODOLITE RECORDS OF WIND SPEED, DIRECTION AND HEIGHT

the receiver is $6 \cdot 25 \times 10^{-6}/40$, i.e. $1 \cdot 5 \times 10^{-7}$ W., or about half the calculated power at the maximum range. In practice it is found that there is ample power available to trigger the receiver at extreme range.

In the case of the air-to-ground (10·5 cm.) link, with a transmitter power of 30 W., transmitter aerial gain about 1, and ground receiver aerial gain (after allowing for a 3 dB. loss due to the offset of the nutating dipole and a further 3 dB. for miscellaneous losses) of 330, the power received at the maximum range is $1 \cdot 9 \times 10^{-10}$ W. The noise power from the radiation resistance of the receiving aerial is given by kTB, where k is Boltzmann's constant, T the effective absolute aerial temperature and B the receiver band width. In the present case this aerial noise power is $8 \cdot 3 \times 10^{-15}$ W. and since the receiver has an overall noise factor of 12 dB. the available noise power is $15 \cdot 85 \times 8 \cdot 3 \times 10^{-15}$ W., i.e. $1 \cdot 3 \times 10^{-13}$ W. The theoretical signal-to-noise ratio is therefore 145 : 1 in terms of power and 12 : 1 in terms of voltage. In practice signal fluctuations and other factors bring the mean signal level appreciably below the theoretical value.

A typical record of a sounding with the prototype equipment is reproduced in Plate XX. It shows the wind speed, direction and height on a strip chart with a scale of 28 cm. wide ruled with 100 divisions, each subdivided in half. One scale division is equivalent to 250 m. in height, 4° in wind direction and 4 kt. in wind speed. Readings can readily be made to one quarter of these quantities. The rulings on the time scale are at intervals equivalent to 50 sec.

5.3. OTHER SECONDARY RADAR SYSTEMS

5.3.1. Lugeon–Nobile system

The principle of this system, which was designed in Switzerland in 1940 and described by Lugeon[56], is illustrated in Fig. 14. The ground transmitter radiated from a dipole aerial a continuous wave on a frequency of 30 Mc./s. with an amplitude modulation at 6 kc./s. This transmission, which had a power of several hundred watts, was detected in an airborne unit, known as a " telesonde ", by a receiver controlling the operation of a transmitter. The 6 kc./s. modulation was applied to the emission of this transmitter, which was on a frequency of 100 Mc./s. and had a power of about 1 W.

The ground equipment had two receivers, one for the signal returned from the airborne unit and the other receiving the signal direct from the ground transmitter.

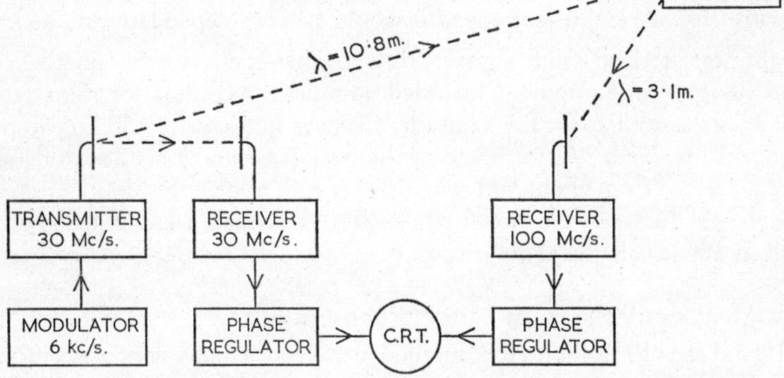

FIG. 14—LUGEON-NOBILE CONTINUOUS WAVE SYSTEM OF RANGE MEASUREMENT

These signals differed in the phase of the 6 kc./s. modulator, the difference being proportional to the travel time of the indirect signal, and therefore to the distance of the airborne unit from the ground equipment. Indication of the phase difference was obtained by feeding one signal through a phase regulator to the horizontal plates of a cathode ray oscilloscope, and the other through a second phase regulator to the vertical plates, thus producing a Lissajous figure on the oscilloscope screen. By adjusting one of the phase regulators until the figure became a straight line, indicating a difference in phase angle of π (or a multiple of π), a measure of the distance of the " telesonde " was obtained in terms of the amount of phase shift inserted by the regulator. The other regulator served to compensate for any phase delay arising in the circuits. With the 6 kc./s. modulator a phase delay of 2π corresponded to a range of 25 km., and an accuracy of setting of the phase regulator to 1° corresponded to a range accuracy of 69 m. and did not change when, with long ranges, the delay exceeded 2π. The apparatus was reported to operate satisfactorily over ranges of the order of 100 km. For upper wind measurements the system was combined with radio direction-finding measurements of the azimuth of the airborne unit and with height measurements derived from radio-sonde observations.

5.3.2. The " Fledermaus " system

This system was developed in Germany for mobile upper air units during the Second World War and was described by Muller[57] in 1944. The ground equipment, carried in a motor van, was called the " Fledermaus " (bat) and the airborne transponder the " Mucke " (mosquito). In principle the system is similar to that of the Lugeon–Nobile system, the main difference being that the ground-to-air transmission was on the higher frequency (290 Mc./s.) and the return link on the lower frequency (27·7 Mc./s.). The modulation frequency was 7·5 kc./s. and so a phase-angle shift of 2π represented a range of 20 km. Phase comparison was made fully automatic by arranging for a measuring-bridge circuit to control, through a servo-motor, the phase-shift regulator. This automatic adjustment could be made to within a fraction of a degree, giving an accuracy of about $\pm$ 50 m. in range.

Azimuth measurement was made by using the directional characteristics of the ground transmitter aerial, the field-strength polar diagram of which had two strong lobes and a sharply defined minimum between them. One of the lobes was used for range measurements but once every minute the aerial was rotated through 360° and azimuth recorded as a sharp dip in an otherwise steady trace, the dip corresponding to the alignment of the field-strength minimum with the direction of the transponder. Azimuth accuracy was reported to vary from about 0·1° to 0·5°: with the wavelength employed it would be very dependent on site errors.

Height was derived from signals from a radio-sonde carried in the airborne unit and the ground equipment included automatic recorders for slant range and the sonde data as well as for the azimuth. Ranges in excess of 150 km. were stated to have been obtained. An estimate of the overall accuracy of the wind measurements was derived from about 500 comparisons with optical theodolite observations. The differences between the two methods were less than 4 kt. in speed and 5° in direction in about half the comparisons.

5.3.3. Lugeon's pulse system—the " Echosonde "

In 1953 Lugeon[58] proposed a method in which a radio-sonde performed the function of a transponder. In this system the ground transmitter radiates pulses of the order of a microsecond's duration at a fixed recurrence rate and these are received by the transmitting aerial of the radio-sonde, which is itself designed to emit pulses with a recurrence rate near that of the pulses from the ground. Between consecutive pulses the radio-sonde transmitter is quiescent and towards the end of this period it acts as a receiver. It then receives the pulse from the ground which automatically initiates the return pulse from the radio-sonde aerial. The sonde circuit is that of a normal transmitter except that the resistance and capacitance of the grid of the oscillator are such as to produce the self-quenching action. A cathode ray tube is used to measure the time interval between the outgoing and return pulses. The system has been tried out on various wavelengths but it does not yet appear to have been adopted for routine use.

5.3.4. French radar transponder

A transponder developed for use with the radar equipment of French ocean weather ships has been described by l'Hermitte[59]. The radar on these ships has two aerial systems, one of which is for horizontally polarized radiation and the other for vertically polarized radiation at a slightly different frequency. This arrangement provides a means of separating signals to and from a transponder.

The transponder has two double-triode valves, one of which is used as a super-regenerative receiver with one of its triodes acting as a self-quenching oscillator. With a suitably disposed aerial this valve receives the horizontally polarized radar pulse transmission on a frequency of about 220 Mc./s. The output from the receiver is fed into one triode of the other valve, the second triode of which acts as a high-frequency oscillator for the transmission of return pulses. These are emitted from a vertically polarized aerial on a frequency of about 180 Mc./s. A barometrically controlled switch is fitted to the transponder to provide pressure indications by breaking contacts at pre-determined pressure intervals. The power of the transponder is about 4 W. and ranges well in excess of 100 km. are reported to have been obtained in soundings from weather ships.

CHAPTER 6

RADIO DIRECTION-FINDERS FOR METEOROLOGICAL USE

6.1. INTRODUCTION

The two main applications of radio direction-finding technique in meteorology are for upper wind measurement and for thunderstorm location. Since the latter application involves only ground equipment it is questionable whether it should be regarded as an upper air measurement, but as the equipment is similar in principle to the radio direction-finders used for upper wind measurement it is appropriate to include a description of it in this chapter. The application of radar to storm detection is dealt with in Chapter 11. The main difference between radio and radar techniques for this purpose is that the former is capable of giving useful (directional) measurements to ranges of thousands of kilometres, whereas storm warning radars (giving range as well as direction) are at present restricted to maximum ranges of a few hundred kilometres. This is because radar detection depends on the echoes produced by precipitation, whereas in radio detection it is the electromagnetic radiation generated by lightning flashes which produces the signals in radio reception known as atmospherics or, in their meteorological application, as sferics.

The main difference between radio and radar methods of tracking balloons for wind-finding is that in the former the signal is not controlled from the ground but is radiated from a transmitter carried by the balloon and not from a transponder or a reflector. There is, therefore, no facility for measuring range by timing the passage of the signal and the only way of obtaining range with the radio direction-finding technique is by triangulation, as in double optical theodolite ascents.

Direction-finding by radio is essentially the observation of the arrival of a signal, using either the phase difference of the alternating voltages induced in a pair of receiving aerials, or the voltage round a closed loop induced by the oscillating magnetic flux through it. The aerial system may be fixed and consist of two pairs of aerials or two coils in perpendicular planes, the direction being measured by comparing the voltages with a radio-goniometer. Alternatively, it may consist of a single pair or coil which can be rotated and, by noting the changes of signal strength, set in the plane of the wave front, which is the position of minimum response. In the early application to wind-finding, such as that adopted by the Meteorological Office shortly before the Second World War, the measurements were confined to those of the bearing of a balloon-borne transmitter, it being difficult at the time to design a direction-finder to measure angles of elevation. It was necessary, therefore, to make simultaneous observations from two stations to obtain the plan-positions of the balloon (from the points of intersection of the bearings) and also, in order to avoid having to assume a constant rate of ascent, to arrange for the transmitter to signal pressure (and preferably temperature as well) so that the wind measurements could be related to known levels. For direction-finding the signal must be continuous and of constant radio-frequency and some types of radio-sonde, such as those employing a code-type signal (see Section 7.2.3), are unsuitable.

Further development of radio direction-finding led to the introduction, at first mainly in the United States of America, of equipments with which the angle of elevation as well as the azimuth of the balloon-borne transmitter could be measured. Such equipments are generally known as radio-theodolites and they are usually combined with radio-sonde equipment. At present at least three types of radio-theodolite are available. One of these, known as Metox, is a French-made instrument similar in design to the United States' wartime equipment SCR 658. The second is the United States' automatic equipment known as rawin set AN/GMD–1 (and later models) and the third is a system, designed by Väisälä, which differs from the others in having pairs of fixed aerials instead of a single rotatable aerial.

6.2. METEOROLOGICAL OFFICE RADIO-WIND DIRECTION-FINDING SYSTEM

Although this system is obsolete its highly successful use for the important operation of wind-finding up to long ranges during the Second World War, together with the fact that the Meteorological Office Handbook describing it is out of print, justifies a brief record here of its main features. A general description of the system has been given by Harrison[31] and details of the design of the direction-finders by Hopkins[60]. At each station three direction-finder sites were set up, with base lines of length between 30 and 75 km. so situated as to provide adequate triangulation for all wind directions. The type of direction-finder used is shown in Plate XXI. A sensitive superheterodyne receiver was used to detect the signals, at a frequency of about 25 Mc./s. for a balloon-borne transmitter, a frequency changer being used to amplify and convert the signal to an intermediate frequency of 2 Mc./s. After further amplification followed by a second detector stage the signal passed through an audio-frequency output stage to the headphones of the operator. The aerial system used was of the Adcock type, with two vertical dipoles 2 m. apart, mounted on the frequency changer unit which was supported on the rotor of a circular observing table, connexion to the combined intermediate and audio-frequency unit at floor level being by means of slip rings. The orientation of the rotor, and therefore of the aerial system, was observed by either of two pointers against a large, stationary, annular scale divided in half-degrees.

When such an aerial system is orientated in the plane of the wave front of a vertically polarized signal and normal to the direction of propagation, the potential difference at the receiver terminals is zero since the voltages induced in the two vertical aerials are in phase. When the aerial system is rotated from the plane of the wave front the voltages are no longer in phase and an oscillating potential difference is applied to the receiver. Ideally, the minimum signal strength is zero, and a very small rotation of the aerial system is sufficient to produce an audible signal, so enabling the direction to be determined with a high degree of accuracy. In practice this was not always achieved mainly owing to the effects of secondary signals reflected from objects on the ground; under the best conditions it was possible to determine the wind with a vector error not exceeding 3 kt. over the whole of an ascent, but in general the errors were several times as large as this. The chief disadvantage of the method, however, was the great inconvenience of operating the triangular networks with very long base lines, necessitating expensive telephonic communications and transport arrangements. With the advent of radar this problem disappeared and the method became obsolescent, but until recently it retained the advantage of being capable of covering longer ranges than were practicable with primary radar.

6.3. THE METOX RADIO-THEODOLITE

The design of the Metox radio-theodolite follows very closely that of the SCR 658 set, a description of which has been given by Kirkman and LeBedda[36]. A general view of the radio-theodolite is shown in Plate XXII. The equipment is a transportable radio-wind direction-finder primarily designed for tracking a balloon-borne transmitter but also capable, if the latter is frequency-modulated, of receiving radio-sonde information. The ground equipment consists principally of a directional receiving aerial system with a lobe-switching assembly, a receiver incorporating a cathode ray tube indicator, and a power supply unit. The whole apparatus, apart from the power unit, is mounted on an adjustable tripod pedestal. It weighs 430 kg. and requires a power input of 400 W.

The aerial array is in the form of a plane square framework divided into four bays each composed of eight half-wave elements spaced a quarter of a wavelength apart. It is mounted so that it can be rotated in azimuth and elevation. A motor-driven switch selects pairs of the bays in rapid succession and connects them to the receiver. The rapid sequence of lobe displacement up, down, right and left is indicated by pairs of deflexions on a cathode ray tube trace and when each pair, corresponding to horizontal and vertical lobe shifts, is properly matched in amplitude the aerial is orientated in the direction of the airborne transmitter. Sweep voltages for the cathode ray tube are synchronized by contacts associated with the aerial switch mechanism. Movement of the aerial array is controlled by handwheels and azimuth and elevation are read from dials geared to the handwheels. The operator is accommodated on a seat attached to the pedestal.

The receiver, which covers a frequency range of 390 to 407 Mc./s., comprises a radio-frequency amplifier, a mixer, four stages of intermediate-frequency amplification, a detector and a low-frequency amplifier. The direction-finding signal reaches the receiver as a carrier wave with variations in amplitude caused by the lobe-switching cycle when the aerial is " off target ", the amplitude at each contact of the aerial switch being a function of the signal strength received from a particular lobe position.

When the radio-theodolite is used for wind-finding only, and not in conjunction with a radio-sonde, the airborne unit is fitted with a baroswitch to signal pressure steps which are used for determining the altitude.

The incremental angular accuracy of the Metox equipment is stated to be $\pm 0{\cdot}25°$ for elevation angles above 15°. At lower elevations the effect of ground reflections causes large and variable errors.

A remotely controlled version of the Metox equipment has recently been produced, allowing it to be operated from a distance of several hundred metres. It also incorporates improvements in the circuits to allow it to receive transmissions from a Metox pulse-modulated radio-sonde.

6.4. U.S.A. RADIO-THEODOLITES

6.4.1. Rawin set AN/GMD–1

This set was designed by the United States Signal Corps Engineering Laboratories to give greater accuracy in direction-finding than the SCR 658. It has been described by Kirkman and LeBedda[36] and also by Ference[61]. A block diagram showing the main units of the equipment is shown in Fig. 15. Its principal feature is that it is designed for automatic following of an airborne transmitter radiating on a frequency of 1680 Mc./s., thus providing an increase in basic instrumental accuracy over that of 400 Mc./s. equipment and maintaining that accuracy to a lower angle of elevation. The equipment, therefore, is much more suitable for measurements to great heights and in high winds (i.e. at low angles of elevation) than its predecessor.

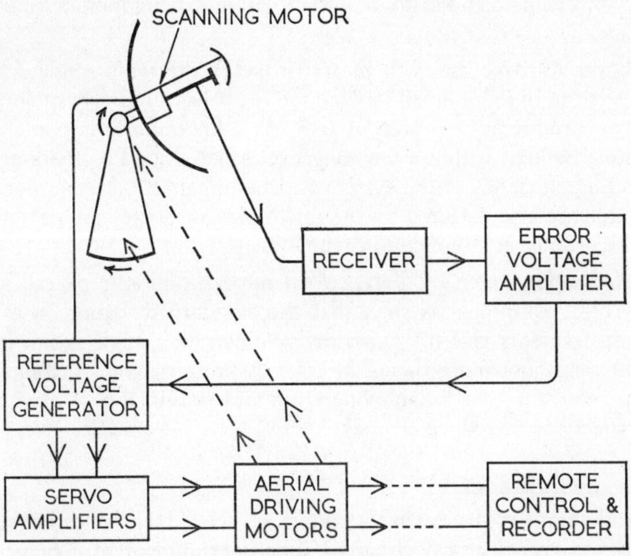

FIG. 15—BLOCK DIAGRAM OF ELECTRICAL UNITS OF RAWIN SET AN/GMD–1

It consists of an aerial system mounted on an adjustable tripod pedestal, mechanism for automatically positioning the aerial, a receiver and a recorder. The total weight is 950 kg. and the units are readily transportable. In the aerial system a stationary half-wave dipole is mounted at the focus of a 7 ft. diameter paraboloid reflector which has a polar diagram with a lobe about 8° wide. A small hemispherical reflector mounted eccentrically in front of the dipole deflects the axis of the lobe from that of the paraboloid by about 3° and rotation of the hemisphere produces a conical scan at a rate of 30 revolutions per second. A signal received on the carrier frequency of 1680 Mc./s. is thus modulated sinusoidally at 30 c./s. and the amplitude and phase of the modulation depend on the amount by which the aerial system is " off target ". The modulated wave passes from the dipole to the receiver unit. Here it beats against the output of a local oscillator to produce an intermediate frequency of 30 Mc./s. on which the modulation and amplitude variations remain. The I.F. signal is amplified and detected and is then used as an error signal in the aerial positioning system.

In the aerial positioning unit the sinusoidal signal from the receiver is amplified and compared with two reference voltages from a voltage generator driven by the scanning motor. These voltages correspond to azimuth and elevation components of the direction of the axis of the aerial system and the comparison with the signal produces two d.c. voltages which are proportional to the amounts in azimuth and elevation by which the aerial is " off target ". The error voltages are applied through amplifiers to driving motors automatically controlling the movement of the aerial system in azimuth and elevation. Thus the axis of the aerial is constantly directed on to the airborne transmitter. Error voltages can also be applied for tracking the transmitter manually.

Synchronous repeaters transmit the azimuth and elevation angles to a remote recorder which indicates and prints the angles at regular time intervals. The aerial control mechanism and azimuth and elevation driving motors can be operated remotely from the recorder unit.

The airborne unit for use with the rawin set is the radio-sonde AN/AMT–4, which is referred to in more detail in Chapter 7. It has a radio-frequency oscillator of cavity type producing a power of 0·5 W. An important point is that the frequency must be kept within a few megacycles per second of 1680 otherwise the direction-finding efficiency of the system will be impaired. Meteorological signals from this sonde are also detected by the rawin receiver; they are passed to special circuits for recording on a separate instrument.

From information given by Ference[61] it appears that the overall accuracy of the AN/GMD–1 equipment is such that the standard deviation in azimuth and elevation measurements is 0·07°, provided the elevation angle is not less than 6° and the slant range not more than 200 km. At longer ranges the signal-to-noise ratio becomes too small for reliable operation and presumably at elevations below 6° ground reflections introduce errors.

6.4.2. Later rawin models

Later productions of the rawin set known as AN/GMD–1A and 1B incorporate improvements in the electrical circuits. The development of a range-measuring attachment for converting the 1B model into a secondary radar system has been described by Todd and Peterson[62]. This equipment is known as AN/GMD–2 and for the range measurement it employs the principle of phase comparison of sinusoidal outgoing and return signals, similar to that described in Section 5.3.

6.4.3. United States Weather Bureau radio-theodolite WBRT–57

This equipment is similar in principle to the GMD–1 but differs from it in a number of features, including the electronic and servo-mechanisms. The set is not designed to be transportable and the mounting of the automatic following aerial system, which is shown in Plate XXIII, is such as to give a high degree of stability and rigidity. The paraboloid is 10 ft. in diameter and provides a stronger signal than the GMD–1. Transmission is on a frequency of 403 Mc./s. and reception on 1680 Mc/s. A photographic telescope is provided on the aerial system and the position of the target from a grid in the picture can be used for checking the accuracy of tracking. This accuracy is expected to be such that the root mean square angular errors do not exceed 0·03°. The set is designed for remote operation from a control unit at a distance up to 2,000 ft. and there is provision for manual control of the aerial system during release of the radio-sonde. Azimuth, elevation and time are automatically printed by the recording unit.

6.5. THE VÄISÄLÄ RADIO-THEODOLITE

This direction-finder was developed by Väisälä[63] for use with the transmitter of the Finnish radio-sonde. As this sonde transmits on the relatively long wavelength of 12·5 m. the use of a direction-finder with a rotatable aerial would not give sufficiently accurate indications. The scheme adopted by Väisälä is the

RADIO DIRECTION-FINDER FORMERLY USED IN THE METEOROLOGICAL OFFICE FOR WIND-FINDING

(79967)

Plate XXII

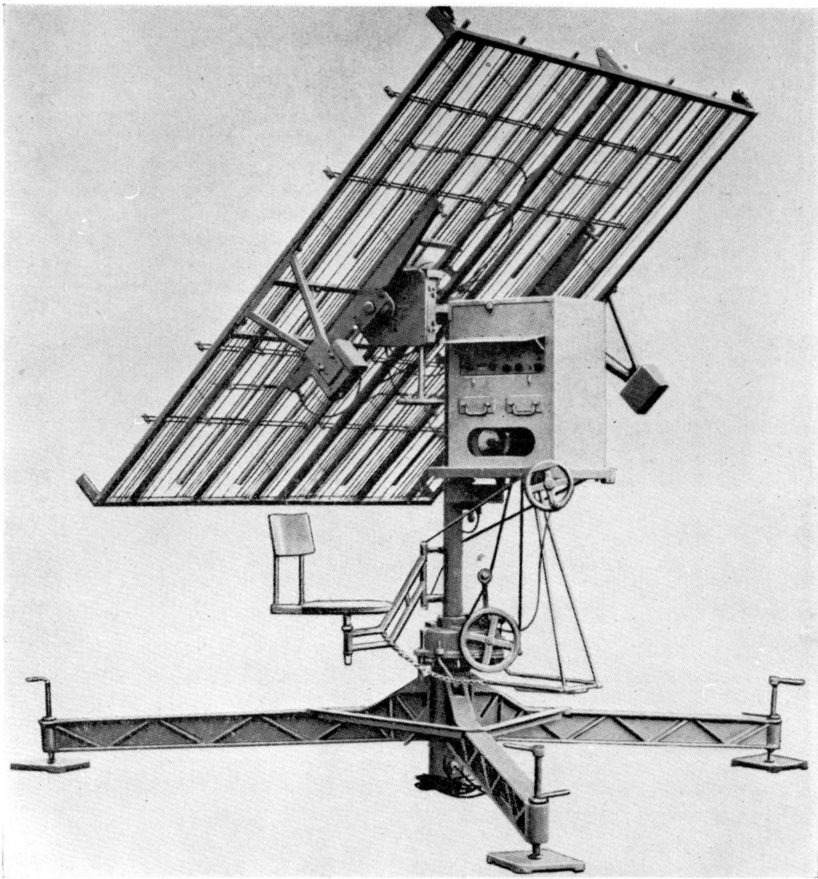

By courtesy of Bendix Aviation Corporation

GENERAL VIEW OF METOX RADIO-THEODOLITE

By courtesy of U.S. Weather Bureau

UNITED STATES WEATHER BUREAU RADIO-THEODOLITE WBRT–57

Plate XXIV

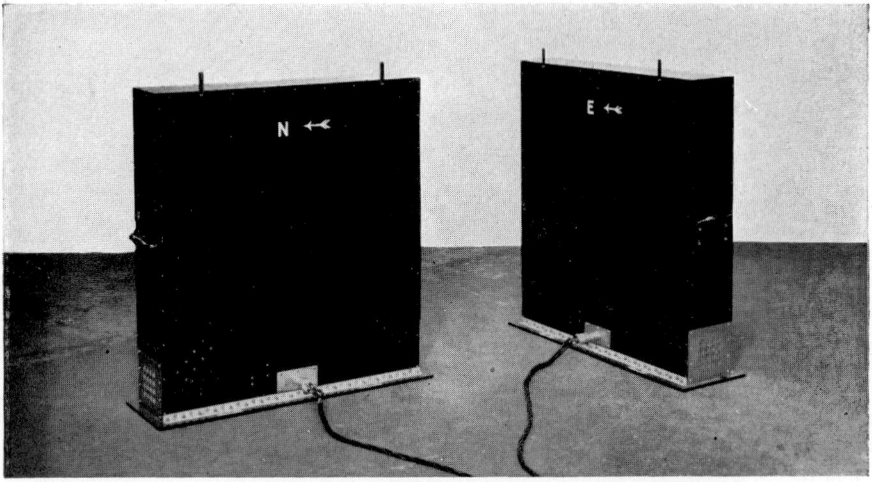

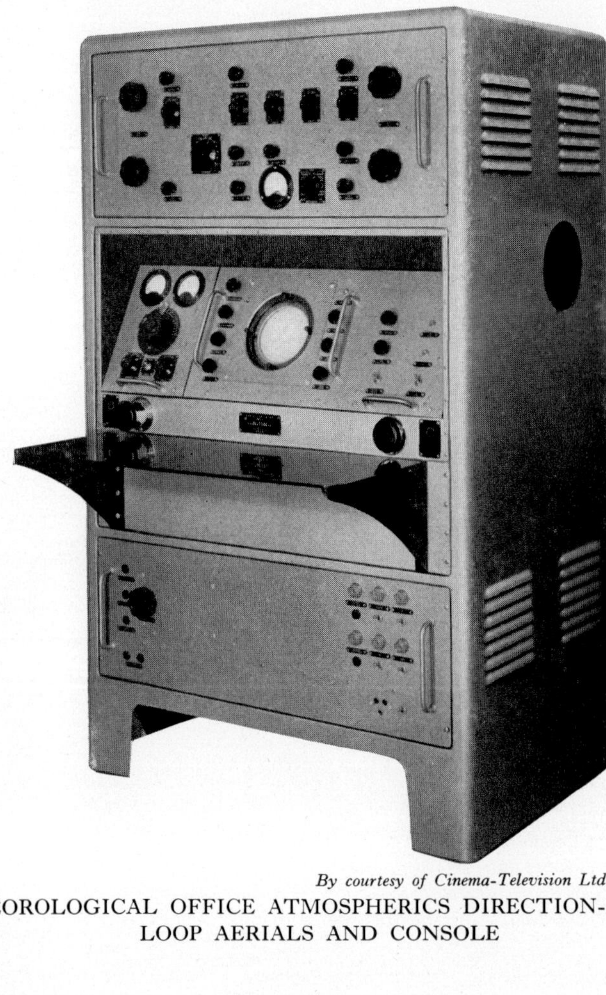

By courtesy of Cinema-Television Ltd.

METEOROLOGICAL OFFICE ATMOSPHERICS DIRECTION-FINDER: LOOP AERIALS AND CONSOLE

measurement of phase difference between waves received by two fixed aerials spaced a few wavelengths apart. The principle of the system is illustrated in Fig. 16.

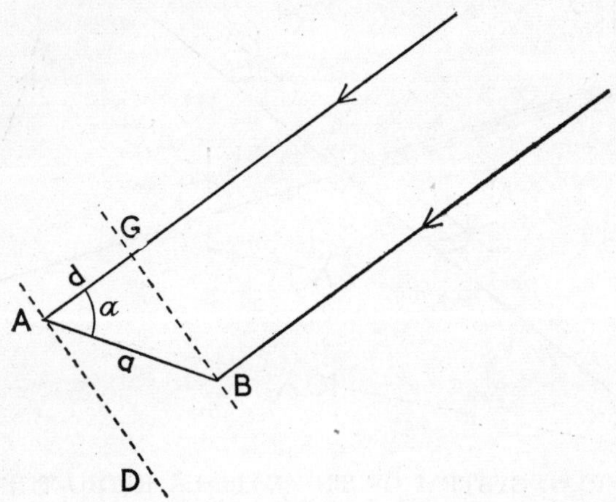

FIG. 16—PRINCIPLE OF THE VÄISÄLÄ RADIO-THEODOLITE

When the airborne transmitter is at a distance that is great relative to the separation a between the two aerials A and B the wave surface approaching them can be considered as plane. The phase difference between the wave fronts GB and AD reaching A and B at the same instant will depend on the distance, d, or $a \cos \alpha$, where α is the angle between the direction of the incoming signal and the line joining the aerials. The measurement is made in terms of the wavelength λ and if d is expressed as $N\lambda$, N being the phase difference, then

$$\cos \alpha = \frac{d}{a} = \frac{N\lambda}{a}.$$

In order to obtain the direction of the signal it is necessary to have two measurements of the angle α with respect to two fixed directions. This necessitates a second pair of aerials and it simplifies the computations if both pairs are the same distance apart and if one pair is on a north–south line (x co-ordinate) and the other on an east–west line (y co-ordinate), as indicated by AB and BC in Fig. 17. We then have:

$$\cos \alpha = \frac{N_x\lambda}{a} \qquad \text{and } \cos \beta = \frac{N_y\lambda}{a}.$$

These angles are related to the azimuth and elevation thus:

$$\cos \alpha = \cos E \cos A \quad \text{and } \cos \beta = \cos E \sin A,$$

from which we find:

$$\cos E = \sqrt{(\cos^2 \alpha + \cos^2 \beta)} \qquad = \frac{\lambda}{a} \sqrt{(N_x^2 + N_y^2)},$$

$$\text{and } \tan A = \frac{\cos \beta}{\cos \alpha} = \frac{N_y}{N_x}.$$

The azimuth and elevation can therefore be obtained from measurements of the phase differences and the wavelength.

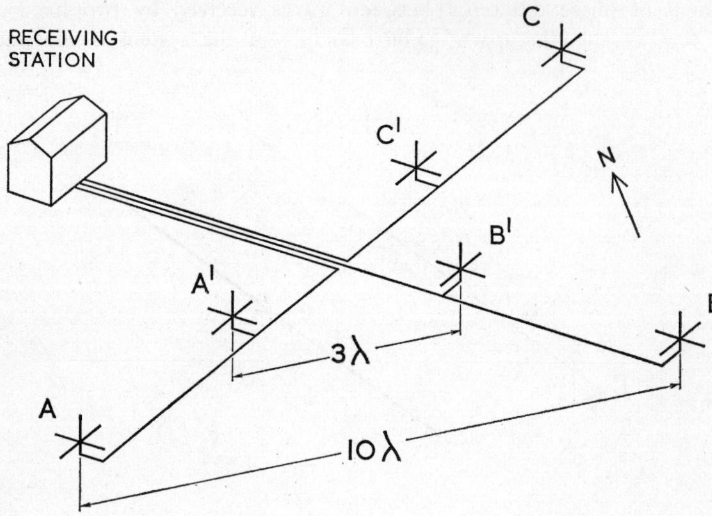

FIG. 17—AERIAL SYSTEM OF THE VÄISÄLÄ RADIO-THEODOLITE

As the Väisälä radio-sonde operates on a variable radio-frequency it is necessary to record the wavelength simultaneously with the phase differences. The spacing a between the aerials is 120 m., i.e. about ten wavelengths, and since N_x and N_y can vary between $\pm$ 10 ambiguity arises as to the number of whole periods in the phase difference. This is determined by using a subsidiary aerial system, indicated by A′B′C′ in Fig. 17, in which the aerials are two or three wavelengths apart, so that the phase difference is less likely to exceed one whole period.

All the aerials are similar and consist of a vertical quarter-wave dipole with a counterpoise in the form of two horizontal half-wave dipoles crossing each other at right-angles. They must be sited on reasonably level and open ground and are connected to the receiver by coaxial cables. The receiver is a twin-channel cathode ray direction-finder incorporating two identical, superheterodyne-type amplifying channels with common local and beat oscillators and a phase-comparison bridge. The phase differences are displayed on the cathode ray tube screen by a linear trace which turns about the centre of the screen through an angle proportional to the phase difference. One operator is needed to operate the receiver but there is provision for semi-automatic continuous recording of the phase difference, the frequency of the radio-sonde signal, and time marks at minute intervals.

The system is said to give an incremental accuracy of $\pm$ 0·1° for all azimuth angles and for an elevation angle of 45°, and $\pm$ 0·5° for an elevation of 11°, below which the accuracy falls off rapidly. Two comparisons with optical theodolites reported by du Toit[64] indicated that differences in winds ranging from 10 to 40 kt. up to 300 mb. were of the order of 6° in direction and 3 kt. in speed.

6.6. RADIO EQUIPMENT FOR THUNDERSTORM LOCATION

6.6.1. General

There are several different types of equipment in use for the detection or measurement of atmospherics, but the type in most common use is the radio direction-finder which gives the direction of arrival of the sferics signal. Two or more stations with direction-finders are required to locate the source of the signal but the possibility of estimating its distance from a single station, for example by analysing the wave form of the received signal, is receiving some attention.

There are two main forms of direction-finder: the cathode ray tube type and the narrow-sector recorder. In the former the signals are received by two similar aerials suitably orientated, and after amplification are applied to the plates of a cathode ray tube. Equipment of this type, which originated with Watson Watt and Herd[33], has been designed and is in use in a number of countries including the United Kingdom, U.S.A., France, Germany, Japan and Sweden; a description of the British equipment is given in the next section. The narrow-sector recorder is a radio direction-finder in which atmospherics from a limited sector are recorded. The sector is related to the orientation of the aerial, which is rotated continuously, and the bearings of the atmospherics are recorded automatically. French and Swiss designs of narrow-sector recorder have been developed.

Other types of equipment in use in some countries include (*a*) rate-of-occurrence meters, providing a record of the total number of atmospherics above a certain signal strength received in a given time interval, (*b*) intensity meters which take into account the individual intensities of the atmospherics received, (*c*) local lightning flash counters which count the number of flashes occurring within a limited radius of the station, and (*d*) wave-form recorders. A survey of all the techniques is given in Chapter 14 of *W.M.O. Publication No.* 8[65].

6.6.2. Meteorological Office atmospherics direction-finder

A general account of the Meteorological Office organization for sferics observations has been given by Ockenden[66]. The equipment now in use is the atmospherics direction-finder, Mark 3, which was designed by the Radio Research Organization and described by Horner[67]. It differs from earlier models described by Adcock and Clarke[68] in being smaller and simpler and more convenient for operation and servicing. It consists of two enclosed loop aerials and a console housing the receiver and associated units, as shown in Plate XXIV. The loops, which are wound in two layers, are about 1 m. square and are orientated in the north–south and east–west directions. They are connected by coaxial cable to twin radio-frequency amplifiers which are normally tuned to a frequency of 10 kc./s. Around this frequency the energy of the spectrum of an atmospheric is relatively high and there is very little interference from man-made signals or other sources. Provision is made, however, for tuning in other pre-determined frequencies by means of switched capacitors.

Each amplifier consists of a cathode-follower input stage, two resistance–capacitance coupled amplifier stages, a phase-splitter and a push–pull output stage. Only the aerial circuits and the output transformers are tuned, the variable capacitors in these circuits serving mainly to adjust the phase response. Adjustment of the amplifier gains is made by means of a pair of matched attenuators.

The signals from the amplifiers are applied to the deflexion plates of a cathode ray tube, the screen of which is displayed in the middle section of the console. They produce on the screen a diametral trace representing the resultant of the signal components in the planes of the aerial loops and therefore with a direction corresponding to the direction of arrival of the atmospheric (but with an ambiguity of 180°). A perspex cursor round the circumference of the screen enables the bearing to be read accurately. There is also a brilliance-modulator unit for momentarily increasing the brilliance of the trace; it is actuated by atmospherics with amplitudes exceeding a pre-determined level.

As atmospherics are of a transient nature, lasting only one or two microseconds and having amplitudes varying in sign and amount, it is essential that the ratio of the components of the signals should be maintained constant throughout the two circuits and that the components should reach the cathode ray tube in exactly the same phase. This is achieved by careful design of the receiver, which provides linear amplification over a very wide range of signal strengths, and by regularly matching each stage of amplification of the receiver. For this purpose a test-signal generator is provided; the test signals are injected into the aerials and corresponding stages in the two amplifiers are adjusted to have the same gain and phase-shift characteristics over a small band of frequencies. These adjustments are carried out stage by stage, switches between corresponding stages of the two amplifiers being provided to enable the grids of the valves to be connected or either one earthed.

The console of the direction-finder incorporates built-in telephone equipment and an automatic selector for use in the co-ordination of observations at the stations in the network. At any one station each atmospheric of sufficient intensity to trigger the brilliance modulator produces an audible pulse in the telephone system which serves as a signal for observers at the other stations to read the bearing. The pulse is followed by a quiescent period long enough for the bearings to be recorded. A sense channel, used with a short vertical aerial, is available for resolving the 180° ambiguity in the bearings but it is not normally required when observations from two or more stations are plotted. Further technical details about the direction-finder and its operation are given in the instruction manual supplied with the set.

An assessment of the accuracy of the location of sources of atmospherics by the system of radio direction-finding has been made by Horner[69]. He estimated that with the present British network and observation technique the probable error in position of a storm centre at a distance of 1000 km. is about 20 km. by day in summer, 50 km. in winter and 100 km. by night. There are several causes of error. Site errors may be present if surrounding objects are large enough and near enough to reflect sufficient energy to deviate the displayed bearing from its true value. Another source of error is the change of polarization of waves reflected from the ionosphere. An equipment error occurs when a wanted signal arrives before the oscillations due to a preceding wave have died out. Finally there are random errors of reading and plotting.

6.6.3. Narrow-sector recorders

In these equipments the aerial system consists of two loop aerials mounted at right-angles to each other about a common vertical axis round which they rotate continuously. The loops are connected to two amplifiers, one of which has its output also connected to a non-directional aerial. Both outputs are fed through a " block relay " to a recorder in such a way that the latter only receives an atmospheric signal when the plane of the loop associated with the amplifier that is also connected to the non-directional aerial is approximately at right-angles to the direction of arrival of the signal. Thus the apparatus is only sensitive to signals from a direction within a sector the width of which depends on the ratio of the gains of the amplifier circuits and is normally adjusted to be about 10°. The signals are recorded on a cylindrical drum which rotates in synchronism with the aerial system and the direction of arrival of an atmospheric is automatically plotted against time. The record is used for obtaining the mean bearing of the source of a number of atmospheric signals rather than for locating individual systems.

6.6.4. Local lightning flash counters

The purpose of these instruments, which are sometimes known as ceraunometers, is to count the number of lightning flashes which occur within a limited radius from the station. They are radio receivers which count or record each atmospheric signal received within a limited range of signal strength. Unfortunately the signal strength of an atmospheric depends not only on the distance from the source but also on the strength of the source and on propagation and polarization conditions and so it is impracticable to achieve a definite limit to the radius of operation. It is important, therefore, that for any network of lightning flash counters a common standard of performance should be adopted; an international standard design of instrument suitable for recording flashes within a radius of about 50 km. has recently been recommended[70].

CHAPTER 7

PRINCIPLES OF RADIO-SONDE DESIGN

7.1. GENERAL CONSIDERATIONS

7.1.1. Basic requirements

The primary function of the conventional radio-sonde is to provide measurements of pressure, temperature and humidity from ground level up to heights of between 20 and 40 km. and to transmit the measurements to a ground station. Radio-sondes have also been designed or adapted to measure other properties of the atmosphere such as the electrical potential gradient, ozone content, solar and sky radiation. Since measurements may be required to be made in winds of all strengths, it is necessary for a sonde to be able to transmit the information over the greatest ranges likely to be covered by a balloon ascending to the required height at the rate normally used. Thus, if the mean wind up to 30 km. is 60 kt. (30 m./sec.) a balloon reaching that height in an ascent at 6 m./sec. will have covered a horizontal range of 150 km. To cover any extreme winds that may be experienced, satisfactory reception of signals over slant ranges of 200 km. is therefore desirable.

The main factors influencing radio-sonde design, apart from the specification of accuracy, are weight, power and freedom from temperature and pressure effects other than those in the measuring elements concerned. Excessive weight adversely affects both the rate of ascent and the maximum height of a sounding. Power supply, which usually contributes substantially to the weight, must be kept down to the minimum that is necessary to operate the sonde for little more than the time required for the highest soundings; normally a period of about two hours is ample. Temperature coefficients of components other than the temperature element itself are liable to cause errors; the effects may be reduced by enclosing the components in a thermally insulating case if this is practicable. As battery voltages vary with temperature, it is important that the measuring circuits should be independent of voltage changes. Another important point about power supply is that operation at low pressure may be limited by breakdown of the high voltage supply across small gaps. A pressurized case could be used to prevent this, but it is better that the design of the sonde should take account of this possibility and avoid the use of gaps that are too small in relation to the voltages used.

The conditions under which radio-sondes often have to be launched require them to be of robust construction. On the other hand, as the majority of them are used only once, they should be designed for mass production at low cost. Ease and stability of calibration is another important factor.

A radio-sonde generally comprises three main parts, viz. the sensing elements, the radio transmitter, and some form of " pick-up " for converting the indications of the sensing elements into electrical variations that modulate or control the transmitter. If, as is normally the case, more than one meteorological quantity is to be measured during each cycle of operations, some means of switching each measuring element in turn into the transmitting circuit is generally necessary. This may be done by driving a switching device by an electric or a clockwork (spring or gravity operated) motor, by a windmill or by an aneroid pressure element.

A windmill drive is inexpensive and is reliable over most of an ascent but, depending as it does on the mass rate of air flow created by the ascending balloon, is liable to fail at high levels.

Most radio-sondes operate on frequency bands between 25 and 100 Mc./s. but some use a band near 400 Mc./s.; all of these usually employ simple transmitters of conventional design. In some recent developments frequencies of 1600 Mc./s. and higher are used and the transmitters incorporate cavity oscillators. High frequencies are necessary if the radio-sondes are also to provide signals for wind-finding.

Ground equipment.—The ground equipment of a radio-sonde station varies with the type of instrument with which it is to be used. The barest essentials are a radio receiver and a means of interpreting the signals in terms of the meteorological quantities measured. Automatic recording is not essential except perhaps for systems in which the signals are measured chronometrically or for measurements in which fine structure is to be investigated; for other systems the question is more a matter of economics. The saving in operating staff by the use of automatic recording may be offset by the complexity of the apparatus requiring a higher amount or degree of maintenance. However, the advantages of automatic recording will often outweigh the disadvantages.

7.1.2. Meteorological elements

General requirements.—Sensitive elements for radio-sondes fall into two main categories: those in which the sensitivity to variations in the meteorological quantity resides in an electrical property, and those in which the sensitivity is a non-electrical effect. The former have the advantage of being more readily adaptable to the direct control of the radio signals, whereas the latter generally involve the use of mechanical linkages which, besides adding weight, may introduce friction and backlash. Where mechanical linkages are used the sensitive elements must be free to move with the minimum possible friction; flexible spring hinges should therefore be used instead of pivots.

It is desirable that the meteorological elements should be capable of continuous measurement and that their control of the radio signal should cover a wide range, either as a continuous change or in finite steps sufficiently small to provide the accuracy required. This is more important for temperature and humidity measurement than for pressure. Another important consideration is that the elements should be able to hold their calibration for periods during which they may be in store or in transit.

Pressure elements.—In the majority of radio-sondes aneroid capsules are used for measuring pressure, but in recent years considerable attention has been given to the possibilities of hypsometers. These have the advantage of avoiding mechanical linkage since an electrical thermometer may be used for measuring the boiling-point. Another advantage, particularly for high-altitude soundings, is that the sensitivity of a hypsometer increases rapidly as the pressure decreases.

Aneroid capsules are light and robust but may have an appreciable temperature coefficient and suffer from hysteresis. Their sensitivity depends mainly on the effective surface area and the thickness and elasticity of the material. By the use of metal, such as steel or beryllium copper, with suitable elastic properties, the use of a separate control spring, such as is provided in ordinary aneroid barometers, is unnecessary. An alloy known as nispan has the additional advantage of a very low temperature coefficient. Aneroid capsules are liable to change their sensitivity in the course of time, but this tendency is reduced by a process of seasoning such as 15 to 20 cyclic changes of pressure followed by 5 temperature cycles over the full working ranges. Hysteresis is not a serious defect when the changes of pressure measured are always in the same direction.

The temperature sensitivity of an aneroid capsule depends partly on the temperature coefficient of its elastic constant, partly on the change of pressure of any residual gas, and to a smaller extent on the coefficient of expansion. The latter can, in fact, be used to compensate for the thermo-elastic effect, but only at one particular pressure called the compensation pressure. The method of compensation, which is more useful for ordinary surface aneroid barometers, is discussed in detail in Section 2.3.4 of Part I of this handbook. In radio-sondes it is more usual to employ a bimetal strip to provide compensation for temperature effects, but there is little advantage in this if other parts of the pressure unit, such as an inductor, are temperature sensitive and are not compensated. In a compensation mechanism the thermal time constants of the different components must not differ too much, in order that significant temperature differences may not arise under working conditions. A system which gives good compensation when temperature changes are slow may not do so when they are rapid. It is desirable therefore that all parts of the compensated pressure unit should be similarly shielded from radiation.

One application of the hypsometer method is referred to in the description of the Dutch radio-sonde given in Section 7.2.4. More recently Conover and Stroud[71] have described a hypsometer which has been designed at the United States Signal Corps Engineering Laboratory for high-altitude radio-sonde flights. It consists of a 10 $cm.^3$ capacity vacuum flask filled with cotton saturated with about 5 $cm.^3$ of carbon bisulphide in which a bent-stem, glass-covered bead thermistor is immersed. The purpose of the cotton is to provide adequate surface for evaporation and to minimize super-heating and " bumping ". The vapour pressure of carbon bisulphide decreases from about 400 mb. at 20°C. to about 2 mb. at − 70°C., and as no heater is provided the range of the hypsometer is restricted to pressures less than 400 mb. In flight an aneroid capsule is used for measuring pressures between the ground and this level and to provide a datum pressure for the hypsometer at about 300 mb. With this arrangement flight tests show the hypsometer to give readings to within 2–5 per cent. of, and generally lower than, the aneroid readings in the range 30–2 mb.

Temperature elements.—The chief requirements for a satisfactory temperature element are robustness, stability of calibration, rapidity of response and freedom from radiation errors. Most of the elements used at present are either of the bimetal type or the electrical resistance type. The latter may be a ceramic semi-conductor, such as a thermistor, or a wire resistor.

The theory of thermometric lag is discussed in Section 3.1.6 of Part I and the point of particular importance in the application to radio-sondes is that since the lag coefficient depends on the mass rate of air flow there is a large decrease in the rate of response of the temperature element with height, as can be seen from the following values for the lag coefficients (defined as the time required to respond to 63 per cent. of a sudden change of temperature) of three typical elements.

Temperature element	Lag coefficient in seconds at 0 km.	15 km.
Bimetal, steel/invar 0·25 mm. thick, 15 mm. long	7	22
Thermistor, 2 mm. diameter, 30 mm. long	5	17
Nickel wire, 0·1 mm. diameter	0·02	0·06

The only way of reducing radiation errors to negligible proportions is to use an element in which the heat exchange by forced convection is large compared with the radiation absorbed. An element in the form of a thin cylinder or fine wire with a low absorption coefficient is suitable in this respect. Practically any other form of element needs a radiation shield. In addition to preventing radiation from falling directly on to the element the shield should also prevent radiation reaching it after multiple reflections, it should not itself absorb so much radiation as to warm the air passing over the element, and it should not reduce appreciably the ventilation of the element. Since some of these requirements are mutually antagonistic, the design of the shield is necessarily in the nature of a compromise.

The metals most often used for bimetal elements are invar and steel or invar and brass. Again some compromise has to be made in the practical design since an element must be sufficiently stiff to control its linkage mechanism and, whereas the stiffness increases with the cube of the thickness of the bimetal, the sensitivity decreases and the lag coefficient increases linearly with the thickness. The stability of bimetal elements is improved by subjecting them to a number of cycles of temperature change over the working range. When properly seasoned such an element shows little hysteresis and is capable of an accuracy of the order of 0·1°C.

Temperature elements of the semi-conductor types are usually made of a ceramic material the resistivity of which changes with temperature. This type, generally known as a thermistor, has a high resistance R which decreases with increase of absolute temperature T roughly in accordance with a relationship of the form:

$$R = Ae^{B/T},$$

where A and B are constants.

As, therefore, the sensitivity is inversely proportional to the square of the absolute temperature the scale is very far from linear. Thermistors can be made very small and thus have small lag and radiation errors, but their power dissipation must be kept low to prevent self-heating. They are not as stable in performance as metal wire resistors and are not so readily reproducible for sensitivity.

Elements in which the permittivity varies with temperature may be used to control signals by changes in capacitance. They employ a ceramic dielectric, such as barium strontium titanate, which has a suitably high temperature coefficient.

Wire resistors have the advantages of high stability, linearity of scale and low lag and radiation errors. They are liable to be rather fragile since a small diameter is necessary if full advantage of this type of element is to be obtained. The materials generally used are nickel or tungsten.

Humidity elements.—There are two main categories of humidity elements in common use for radio-sondes. One category utilizes the effect of humidity on the dimensions of certain materials such as hair and gold-beater's skin, and the other the effect on electrical resistance of films of hygroscopic substances. The characteristics of hygrometers of both categories are discussed in Chapter 4 of Part I and an up-to-date review of the electrical types has been written by Wexler[72].

In radio-sonde applications the main difficulty arises from the fact that the response of humidity-sensitive materials becomes very sluggish at low temperatures. Whereas at room temperature lag coefficients, expressed as the time required to indicate 63 per cent. of a sudden change in humidity, of a few seconds or tens of seconds are obtained, they may increase to about twenty times as long at −30°C.

Ordinary hair is the worst material in this respect, while gold-beater's skin and some hygroscopic film elements are three or four times as rapid in response as hair. A very great improvement in the response of hairs at low temperature has been achieved, at the expense of roughly half the tensile strength, by Frankenberger[73] by flattening the hairs between rollers under high pressure. Both hair and gold-beater's skin exhibit hysteresis after exposure to low humidities.

The best-known electrical resistance humidity element is that developed by Dunmore[74] for radio-sondes in the U.S.A. This consists of a polystyrene strip coated with a film of lithium chloride dissolved in polyvinyl acetate and having two tin electrodes sputtered on the long edges. The resistance at room temperature ranges from about 6 MΩ to 4000 Ω for a change of relative humidity from 15 to 100 per cent. These elements are also temperature-sensitive and corrections have to be applied. It has been found by Wexler[75] that a very much faster response with an element of the resistance type can be obtained by using a thin film of potassium metaphosphate deposited on a glass slide.

Attempts have been made, e.g. by Smith and Hoeflich[76], to develop an electrical humidity element in which dimensional changes produce changes in resistance. This has been done by suspending carbon particles in a plastic binder. The latter changes its volume with relative humidity and in so doing alters the relative positions of the carbon particles and changes the resistance. This type of hygrometer unit is said to be faster in response and to be less sensitive to temperature than the lithium chloride type, but it does not appear to have been adopted for wide-scale use.

A novel electrical hygrometer described by Taylor[77], and also by Crawshaw and Davidson[78], possesses features which may make it worth-while developing for radio-sonde use. It consists essentially of an electrolytic cell in the form of a tube the inside of which is coated with a film of partially hydrated phosphorus pentoxide. Two platinum wire electrodes are spirally wound in contact with the film. The latter absorbs water vapour from air (or other gases) flowing through the tube and the absorbed water is electrolysed by a d.c. voltage across the electrodes thus producing a current which is a measure of the rate of water absorption, provided the electrolytic process is fully efficient. The method is said to be capable of measuring concentrations of a few parts per million and should therefore be able to determine the very small amounts of water vapour found in the atmosphere at very low temperatures.

7.1.3. Transmitters and power supplies

Transmitters.—In general the design of transmitters for radio-sondes presents no great difficulty and in most of the instruments in current use the transmitter is of a fairly simple type. It must, of course, be light in weight and have a low power consumption. The use of miniature components has greatly facilitated design in these respects, but such components are often costly compared with those of standard size. Stability of operation of the transmitter, with minimum frequency drift due to change in battery voltage and temperature, is an important requirement, especially in systems in which radio- or audio-frequency variation is used as a measure of the meteorological quantities.

For transmitters of conventional design a single triode radio-frequency oscillator with an output from a half-wave or quarter-wave aerial of about 30 mW. will generally provide adequate signal strength at ranges up to about 200 km. The aerial is generally more or less vertical and may form the suspension between the radio-sonde and the balloon. With a vertical aerial the radiation is the same in all azimuths but decreases from a maximum in horizontal directions to zero in the vertical direction. The field strength F (in volts per metre) from a vertical half-wave aerial in free space and carrying a current of I amperes is, in fact, given approximately by:

$$F = \frac{60\, I \cos E}{R}$$

where R is the slant range in metres and E is the angle of elevation (with respect to a plane perpendicular to the axis of the aerial). In practice, however, the signal does not completely fade out if the radio-sonde drifts overhead since the range is not great and the swinging of the instrument on its suspension causes the aerial to swing out of the vertical direction by as much as 30°. In some recent developments, in which requirements for wind-finding are also catered for, radio-sondes have been designed to operate on frequencies of 1600 Mc./s. and higher. These necessitate the use of cavity type oscillators and higher power consumption.

Power supplies.—The chief requirements for the power supply of a radio-sonde are:

(*a*) Small weight.

(*b*) Little or no variation of voltage during discharge.

(*c*) Long shelf-life before use.

(*d*) Relative insensitivity to low temperatures.

The low-tension power for a radio-sonde must be obtained from a battery but the high-tension supply may be obtained either from a battery or from a vibrator driven by the low-tension battery. The design of these items is important since they contribute very much to the weight of the equipment. Three types of battery are in common use, namely, the dry type, the wet acid type and water-activated batteries. Dry batteries, though convenient in use, are relatively heavy and have a limited shelf-life. Lead–acid batteries are too heavy for radio-sonde use but Väisälä[79] introduced a modification of the wet acid type of battery by using a zinc plate as the negative electrode. This not only reduced the weight but produced an appreciably higher voltage than the lead accumulator. The battery can be stored for long periods in the dry state and is made active by the addition of acid shortly before use. The convenience of water-activated batteries is offset, to some extent, by their higher cost. One such battery uses cuprous cloride and magnesium and is said to have a capacity/weight ratio superior to that of lead–acid cells. This type of battery generates sufficient internal heat to make thermal insulation unnecessary. With dry cells and wet acid types the adverse effect of low temperature makes some thermal insulation necessary; failure of such cells may occur at temperatures between −15° and −30°C.

7.1.4. Ground equipment

The design of the ground receiver does not present any difficulty. The two most important features are the sensitivity for a given band width, which governs the minimum signal audible above the electrical noise generated in the circuits and valves, and the selectivity, which determines the ability to separate signals on an adjacent wavelength and to eliminate spurious signals and whistles. The latter quality is associated in a superheterodyne receiver with the requirement for a good signal-to-image rejection factor. With a radiated power of a few tens of milliwatts, representative of many radio-sondes, a receiver sensitivity of the order of a microvolt is required to obtain satisfactory reception at ranges of 200 km. Superheterodyne receivers are generally suitable.

It is a comparatively simple matter to add stages of amplification until the desired sensitivity is reached, but it will be found that a great deal of noise is generated in the frequency changer valve by a periodic fluctuation of emission from the cathode. It is therefore necessary to arrange that when the signal reaches the frequency changer, it is large compared with this noise voltage. This can be effected by the use of a good aerial and by one or more stages of radio-frequency amplification, with valves of high mutual conductance and low noise factor.

The separation of adjacent signals is governed mainly by the number and frequency of the intermediate frequency stages. Trouble is sometimes experienced from image signals; the amplitude of these signals is minimized by the use of radio-frequency tuned circuits before the frequency changer and by the use of a high intermediate frequency.

From the above considerations it is clear that the signal-to-noise ratio is the principal factor determining the satisfactory reception of a weak signal such as that of a distant radio-sonde. It must be strongly emphasized that no radio receiver can compensate adequately for the absence of a good aerial. The latter may be tuned or aperiodic; a ground-plane aerial will generally give the best results.

The other essential items of the ground equipment are those for interpreting the received signals in terms of the meteorological quantities. Since, as already mentioned, these items differ according to the type of radio-sonde system employed, they are dealt with under the various types of system described in Section 7.2.

If the equipment employs automatic recording apparatus this will generally require a stronger signal for efficient operation than does manually operated visual-indicating equipment, but is capable of greater accuracy.

7.2. TYPES OF RADIO-SONDE

7.2.1. General

A distinction may be drawn first of all between radio-sondes that are used for special investigations and those that are used for routine meteorological soundings. The former are generally used in smaller numbers and may be specially designed and constructed (perhaps by hand) to meet the particular requirements involved, for example, in the measurement of such quantities as ozone content or potential gradient. For routine measurements of upper air pressure, temperature and humidity, on the other hand, radio-sondes are used in such large numbers that only those designs that are suitable for mass production are likely to be acceptable. The types of radio-sonde in routine use on a large scale can conveniently be classified according to the method of converting the indications of the sensitive elements into electrical signals that are suitable for radio transmission. The four main methods are as follows:

(*a*) Chronometric,

(*b*) Coding,

(*c*) Variation of radio carrier frequency,

(*d*) Variation of audio modulation frequency.

There are a dozen or so designs in current use employing one or other of these methods and more of them use the chronometric system than any of the other methods. In the following sections on each method the descriptions are confined to one or two examples of each class. Most of the radio-sondes designed up to 1939 are described in a monograph of the International Aerological Commission[26].

7.2.2. Chronometric radio-sondes

The basic principle of the chronometric type of radio-sonde is the same as that introduced by Olland (see von Baumhauer[19]) in 1874 as a method of transmitting instrument readings along a telegraph wire.

It is essentially a time cycle in which the required information is conveyed in terms of a time interval between reference signals, repeated at regular intervals, and signals controlled by the measuring instruments. The application of this principle to radio-sondes consists in using the signals to interrupt, or momentarily alter the frequency of, the carrier wave of the transmitter. In Olland's telegraphic instrument the signals were made by an arm rotating at a constant rate. During each rotation the arm made momentary electrical connexion with one or more fixed reference contacts and with movable contacts, the position of which was controlled by the sensitive measuring elements. There have been diverse variations of this scanning arrangement. One essential requirement in a good design is that the movable members connected to the elements should be quite free except at the moment of the scanning contact. Another important consideration is that the accuracy of the system depends mainly on the short-term constancy of the rate of rotation of the scanning unit. This is difficult to achieve with inexpensive electric driving motors and clockwork motors would probably be better in this respect but for the fact that they produce a discontinuous motion; both types tend to be unreliable at low temperature.

Some idea of the precision that is needed in the construction of a chronometric radio-sonde may be obtained from the following considerations. In order to obtain sufficient detail from a sounding the scanning cycle should not exceed 20 sec., and if a sensitivity of, say, 2 mb. in a range of 1000 mb. is required for the pressure measurement, the scanning speed must be constant to within 1 part in 500 and the time of the pressure contact must be determined to an accuracy of 0·04 sec.

The main features shown by some designs of chronometric radio-sondes are given in the following brief descriptions.

Canadian chronometric radio-sonde[80].—This is one of the simplest designs of this type of sonde. The scanning unit is in the form of an aluminium disc which is maintained in rotation by an electric motor at a rate of about one turn in 13 sec. On the upper surface of the disc a spiral groove is cut, as shown in Fig. 18, and at the circumference two short strips are removed to provide reference marks. The surface of the disc is anodized to make it electrically non-conducting. Movable contact arms controlled by the meteorological elements, and a fixed reference arm, ride on the surface of the disc. Each arm is, in fact, a pair of metal strips which make electrical contact with each other and with the disc only when their tips drop into the spiral groove or, in the case of the reference arm, into the gaps on the circumference. When a contact is made it causes a relay to switch off the high tension supply to the anode of the transmitter valve for a short period of time. The time interval between contacts made by one of the movable arms and the reference arm clearly depends on the position of the movable arm and it is, therefore, a measure of the movement of the sensitive element controlling the arm. The electric motor that drives the scanning disc is provided with a speed control in the form of a vibrating tuned reed which keeps the speed over any one cycle within the required limits of variation.

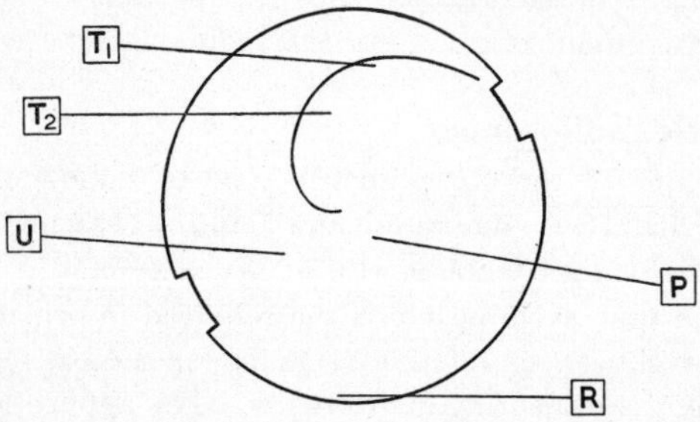

FIG. 18—DESIGN OF SCANNING DISC OF CANADIAN RADIO-SONDE
P, T and U are the variable contact arms and R is the reference contact arm.

The signal received at the ground is fed into a recorder, the chief feature of which is a rotating drum on which is a single-turn helical ridge. The rate of rotation of the drum is automatically controlled to be synchronous with that of the scanning disc in the sonde. A waxed paper chart is carried round with the drum, and a tapper bar actuated by the received signals causes dots to be made on the paper where the bar intersects the helical ridge. The record therefore consists of lines of dots, those corresponding to the reference signals being in a straight line which provides a datum for the measurement of the other lines corresponding to the variations in the meteorological elements.

Bureau radio-sonde.—The special feature of this chronometric radio-sonde which was designed by Bureau[81] about 25 years ago is the method of eliminating the effect of irregularities in scanning speed by the use of timing pulses of sufficient frequency, e.g. 50 per second, to allow the scanning intervals to be measured in terms of the number of such pulses. In a recent model of the sonde the scanning disc is driven by an electric motor which also drives, through gearing, a toothed wheel. The passage of the teeth past a contactor causes a series of pulses to be transmitted. These pulses are suppressed during the period of contact of the movable and fixed arms on the scanning disc and the angular motion of the disc can therefore be measured as a function not of time but of the number of pulses between contacts. The signals are recorded on a tape-chronograph type of recorder and, to facilitate the counting of the pulses, every tenth one is suppressed by the omission of one tooth in ten on the toothed wheel.

The Bureau pulse-counting system is also used in a radio-sonde developed in India[82]. In this sonde the scanning unit is in the form of an insulated cylinder on which is fixed a helix of silver wire. The cylinder is turned by a fan type of windmill.

Swiss radio-sonde.—This instrument, which has been described in detail by Lugeon and his collaborators[83], has a scanning unit in the form of a rotating arm, like the original Olland system, but it is used in a way which avoids all moving mechanical and electrical contacts. The pointers controlled by the meteorological elements terminate in small plates which, when immediately above a similar plate at the end of the rotating arm, act as capacitor plates. The position of the pointers is therefore signalled by a sudden change in capacitance, and thus in the radio-frequency, when the rotating arm, which is driven by clockwork, passes below them. Another novel feature of the Swiss instrument is the use of a very fine bimetallic spiral as a temperature element. The spiral is connected to its pointer by a long thin rod or wire coaxial with the spiral, thus allowing the latter to be exposed well away from the body of the instrument.

7.2.3. Code type of radio-sonde

In this class of radio-sonde, the measurements are essentially discontinuous. The range of variation of each meteorological quantity is divided into small steps and a telegraphic code symbol, such as a Morse letter or number, is allotted to each step and can be transmitted as a signal indicating which step is being measured by the meteorological elements. The code may be repetitive without introducing ambiguity in interpretation. An important feature of this system is that no special ground equipment is required since the signals can be received on an ordinary communications receiver. There may be no advantage in this, however, if the simplicity in the ground equipment is offset by complication in the airborne equipment. The earliest successful radio-sonde, that of Molchanov[23], was of the code type and in more modern form it is still used in the U.S.S.R. The German Graw sonde also uses the code principle.

Molchanov radio-sonde.—A recent model of this radio-sonde, the RZO49, has been described in detail by Khakhalin[84]. The coding mechanism consists of a series of contact studs arranged in arcs in a fixed insulating plate over which the pointer, actuated by a meteorological element, can move. The studs are connected to a windmill-driven multi-point switch in such a way that when the pointer makes contact with any particular stud its position is indicated by the number of dots in a signal produced by switch contacts made in rapid succession. In order to cover the range of variation normally required the code is repeated at appropriate intervals.

West German H50 radio-sonde.—The coding device in this instrument, described in detail by Hinzpeter[85], consists of a grooved aluminium hemi-cylinder on which the code symbols are printed by an anodizing process. The hemi-cylinder is rotated about its axis by an electric motor and the tips of the pointers controlled by the measuring elements slide in the grooves over the code contacts. The latter operate a relay which in turn closes the primary circuit of the high-tension transformer supplying the anode of the transmitter valve. The input from the transformer is obtained from a make-and-break device operated by the electric motor spindle. A number code is used and is repeated nearly five times to give a subdivision of about 500 steps in the range of measurement.

Japanese radio-sonde.—A special feature of this code-type of radio-sonde, which has been described by Isono and Huziwara[86], is that it uses two mercury/thallium-in-glass thermometers. Each mercury thermometer has a contact wire fused through the bore; contact between the mercury thread and the wire occurs at temperatures of about −50°C. for one thermometer and at temperatures of about −30°C. for the other. The lag of these thermometers is similar to that of the bimetal and their precise break-contact temperatures are known. Breaking of the contact in each case modifies the code signals in a recognizable manner. Thus the calibration curve of the bimetal is given a three-point check, one at the ground and two in the air, and any shift can be allowed for during computation of the results. The system therefore avoids any assumption as to calibration shift made on the basis of a ground check or, alternatively, the need for a pre-sounding check at several points over the range.

7.2.4. Variable audio-frequency type

In this type of sonde the radio-frequency carrier wave is modulated by an audio-frequency oscillation that varies with the meteorological quantities. The modulation may be effected by inductors, resistors or capacitors controlled by the meteorological elements. With inductors the air gap between an armature and the core of the inductor in the transmitter circuit may be varied by a mechanical linkage. This arrangement suffers from the disadvantage of the inductors being temperature-sensitive; it therefore involves the application of temperature corrections. For resistance and capacitance control of the audio-frequency modulation electrically sensitive temperature and humidity elements may be used, thus avoiding the disadvantages of mechanical linkage. Pressure units, however, are not so readily adaptable for direct electrical control and a mechanical contact linkage with a resistor is generally used.

Radio-sondes using the variable audio-frequency principle have been developed in the United Kingdom, U.S.A. and Holland. The chief features of the American and Dutch instruments are summarized in the following section and a detailed description of the Meteorological Office radio-sonde is given in Chapter 8.

U.S.A. radio-sonde AN/AMT–4.—Radio-sonde development in the U.S.A. has stemmed mainly from the instrument designed at the National Bureau of Standards by Diamond, Hinman and Dunmore[87, 88]. In their design the audio-frequency variation was effected by resistance changes associated with the meteorological units, and the transmitter, radiating on a carrier frequency of about 70 Mc./s., was amplitude modulated. In a later model, frequency modulation was employed with a carrier frequency of 403 Mc./s. which facilitated the use of the transmitters with radio direction-finders. Further development resulted in the radio-sonde AN/AMT–4, with amplitude modulation of a 1680 Mc./s. carrier, for use with the rawin equipment AN/GMD1 referred to in Chapter 6. A description of this sonde has been given by Ference[61]. The two main units in the instrument are the modulator and the transmitter. The modulator includes a selector switch (baroswitch) operated by an aneroid pressure capsule. This switch has 150 silver strips each separated by an insulated segment and, when the pressure changes, the contact arm of the capsule moves across the strips and segments and switches circuits connected to the temperature or humidity elements into the transmitter input. Each strip and segment corresponds with a pressure that is known from a calibration chart; every fifth strip connects a fixed reference resistance to the transmitter circuit, while all the other strips are associated with the humidity resistor circuit and all the segments with the thermometer resistor circuit. Other sequences of contacts are also used.

The aneroid capsule, which operates over a pressure range from 1000 to 5 mb., is made of nispan C which has a very small temperature coefficient. The temperature element is a thermistor in the form of a thin rod, and is coated with a white lead carbonate pigment which gives it an absorption factor for solar radiation of only 6 per cent. Radiation errors therefore are relatively small and in use the element is fully exposed.

As a humidity-sensitive resistor a polystyrene plate coated with a film of lithium chloride is used.

The radio-sonde transmitter comprises a radio-frequency oscillator of the cavity-resonator type, operating at 1680 Mc./s. and giving a power of 0·5 W., and an electronic modulator for feeding the meteorological information into the carrier wave. The electronic modulator is essentially a blocking oscillator, the repetition rate of which depends on the total resistance in the discharge circuit. As this resistance includes one or other of the variable meteorological unit resistors or one of the reference resistances, the repetition rate varies over a range of from 10 to 200 c./s. and so provides the variable frequency at which the carrier is suppressed, and the measurement of which gives the meteorological information. Power supply for the transmitter is provided by a water-activated cuprous chloride/magnesium battery.

The signal is received at the ground on the rawin automatic tracking system, as described in Chapter 6. The audio-frequency output from the rawin set is fed into a frequency meter which converts the signal to a d.c. voltage, directly proportional to the frequency. This voltage operates a servo-system controlling the pen of a recording potentiometer and a record, which can be converted by the use of calibration data into the meteorological measurements, is obtained.

Dutch radio-sonde.—This variable audio-frequency sonde has been described by Hauer and van Tol[89]. It was designed with a view to avoiding moving mechanisms, such as switches, as far as practicable. The principle adopted is to modulate the carrier wave by three audio-frequency signals simultaneously. This necessitates a separate audio-frequency oscillator for each meteorological unit, but by the use of sub-miniature valves the weight has been kept within reasonable limits. A carrier frequency of 28 Mc./s. is used and the output power is about 100 mW. The three audio-frequency ranges are 1 to 3 kc./s. for pressure, 4 to 8 kc./s. for temperature, and 12 to 25 kc./s. for humidity. They are mixed in a reactance valve and are thus converted, in effect, into capacity variations which impose a frequency modulation on the carrier wave.

The temperature element is a thin rod-shaped thermistor, the variable resistance of which controls one of the audio-frequency oscillators. For the pressure measurement the designers have adopted the hypsometer method. The liquid used is freon and it is contained in a small glass vessel with a narrow tubular opening at the top to allow vapour to escape. A layer of cotton wool or similar material 1 cm. thick is used as heat insulation round the container. The freon boils naturally at all of the combinations of temperature and pressure likely to be encountered in most soundings. A second thermistor is used for measuring the boiling-point of the freon; it operates satisfactorily without any surface covering. The oscillator circuits in which the two thermistors operate are of the three-stage phase-shift type, using one valve each. In order that these phase-shift oscillators should have a very low temperature coefficient of frequency variation, special resistors were developed for use in the circuits. The only mechanical component in the sonde is the humidity unit, which has a gold-beater's skin element controlling the air gap of an inductor in a standard inductance–capacitance type of oscillator.

The accuracy aimed at for the Dutch radio-sonde is 0·5°C. for temperature, 5 mb. for pressure and 5 to 10 per cent. for relative humidity.

7.2.5. Variable radio-frequency radio-sonde

In this type of instrument the meteorological units vary the frequency of the carrier wave radiated by the transmitter. Since the frequency is given by

$$\frac{1}{2\pi\sqrt{(LC)}},$$

the variation may be effected by varying the capacitance C or the inductance L of the transmitter circuit. Capacitance variation has the advantage of lending itself to light-weight construction, since it can be produced by the displacement of one plate of a capacitor. An essential requirement of the method is that unwanted frequency variations, which may arise from battery voltage changes or from temperature effects in components, must be prevented or allowed for. The chief disadvantage of the variable radio-frequency system is that the width of the frequency band necessary to cover the range of the variations in use may be such as to make it difficult to avoid interference in parts of the world where frequency channels are crowded. Moreover, frequency bands for meteorological use are allocated by international agreement and these bands are relatively narrow. One of the earliest designs of radio-sondes, that of Duckert[24, 25], employed the variable radio-frequency principle, but the best-known example of this type of instrument in current use is the Väisälä radio-sonde.

Väisälä radio-sonde.—The main features of this instrument, which was developed in Finland by Väisälä[79], are illustrated in Fig. 19. Each of the meteorological elements is mounted in such a way as to control the position of one plate of an air-spaced capacitor. A switch, driven by a windmill, connects each of these capacitors in turn with the oscillatory circuit of the single-valve transmitter. The elements are arranged so that as the pressure and temperature decrease and humidity increases, the capacitance increases and, therefore, the radio-frequency decreases. A frequency band of 1·6 Mc./s. within the range 23·0 to 26·6 Mc./s. is employed, the usual band being 24·0 to 25·6 Mc./s. In order to provide a standard of comparison for the variable frequencies, two fixed reference capacitors are switched into the circuit during each cycle of the switch. This ensures that errors due to permanent frequency drifts in the oscillatory circuit are eliminated.

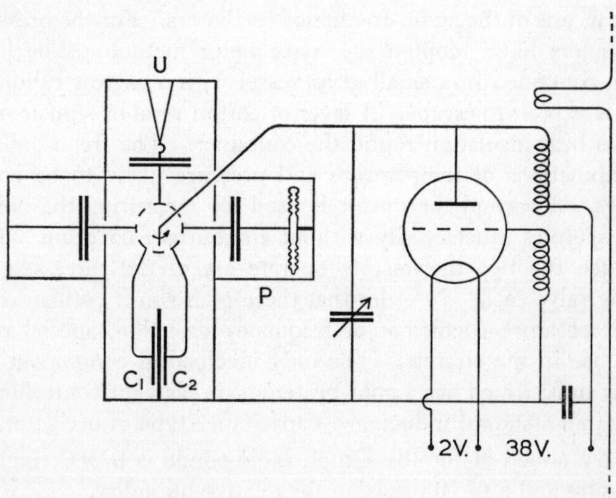

FIG. 19—PRINCIPLE OF THE VÄISÄLÄ RADIO-SONDE

P, T and U are the pressure, temperature and humidity units; C_1 and C_2 are fixed reference capacitors; A is the quarter-wave aerial.

The meteorological elements consist of an aneroid pressure capsule, a bimetallic thermometer and a hair hygrometer. Power is supplied by a lead–zinc battery which delivers 2·4 V. for the valve filament and 38 V. for the anode. The whole sonde is extremely light, weighing only 365 gm. including the battery.

The indications of the meteorological elements are independent of the characteristics of the transmitter, as can be seen from the following considerations. If λ, λ_1 and λ_2 are the wavelengths transmitted when one of the variable capacitors and the two reference capacitors are switched in successively, and if C, C_1 and C_2 are their respective capacitances, then in each case the wavelength is determined by an expression similar to:

$$\lambda^2 = 4\,\pi^2\,L\,(C + C_0)\,,$$

where C_0 is the capacitance of the rest of the transmitter circuit. When L and C_0 are eliminated from the three expressions we have:

$$\frac{\lambda_1{}^2 - \lambda^2}{\lambda_1{}^2 - \lambda_2{}^2} = \frac{C_1 - C}{C_1 - C_2} = y\,.$$

If C_1 and C_2 are constants and C is a function of one of the meteorological quantities, y is a function of this quantity only. It is determined by laboratory calibration of the element in question and since the expression for y does not involve L or C_0 the calibrated curve is independent of the transmitter characteristic.

The wavelengths are measured on a receiver on which the plates of the tuning capacitor are semicircular, so that λ^2 is proportional to the angular readings n. Then,

$$y = \frac{n_1 - n}{n_1 - n_2}\,.$$

The receiver is provided with a semi-automatic recorder which, when a trigger is pressed, registers the readings on a chart on a clockwork-driven drum. The distance between n_1 and n provides a measure of y when $(n_1 - n_2)$ is used as a unit.

Some improvements of the Väisälä radio-sonde have recently been introduced by Rossi[90]. The main modification practically eliminates the effect of the temperature sensitivity of the aneroid capsule. This is achieved by means of a thermostat consisting of a double-walled plastic container filled with water at 0°C. and enclosed in a foam plastic case. In addition to the capsule the transmitter, reference capacitors and switch are housed in the container which, during the fall in temperature in a sounding, is maintained at about 0°C. by the latent heat of ice formation. Other improvements take the form of a bimetallic thermometer element of more rapid response, a more efficient radiation shield and re-positioning of the hygrometer unit to reduce the effect of jolts occurring during launching.

German variable radio-frequency radio-sonde.—An interesting application of the variable radio-frequency principle was made in a radio-sonde developed in Germany during the Second World War. It has been described by Sittel and Menzer[91]. Its chief feature was the elimination of mechanical linkages in the temperature and humidity units, by the use of condensers with a temperature-sensitive dielectric. Two such condensers were used, one for dry-bulb and the other for wet-bulb temperature. For the pressure unit a corrugated diaphragm connected to a cylindrical condenser was used. A small electric motor was used for driving the switch.

7.3. SOME SPECIAL APPLICATIONS OF RADIO-SONDE TECHNIQUE

The vast majority of radio-sondes are used for routine measurements of pressure, temperature and humidity, but there are a number of special applications of radio-sonde technique to other meteorological measurements. These include the use of radio-sondes on balloons floating at constant level, the dropping of radio-sondes on parachutes from aircraft, and measurements of such quantities as atmospheric potential gradient, electrical conductivity, radiation, and ozone content by radio-sonde methods. In order to distinguish radio-sondes that are launched on parachutes from aircraft from those that descend from balloons by parachute, the former are frequently termed drop-sondes.

7.3.1. Constant-level balloon-sondes

The need to obtain upper air information over broad expanses of ocean and inaccessible land areas has led to the development, mainly in the U.S.A., of long-range balloon-borne radio-sondes. For this purpose the balloon, generally of the inextensible type, is controlled to float along a fixed pressure surface for periods of the order of days. The control of the level of the balloon involves the use of a ballast release mechanism and, in the case of a rubber balloon, a gas release valve. By means of a programme timing device the radio-sonde is arranged to transmit its meteorological data at intervals of a few hours and its position at each transmission is determined by radio direction-finding triangulation technique. Wind speed and direction at the level of the flight is therefore obtainable from the trajectory.

There are a number of special features associated with radio-sondes used in this application. In the first place, since the soundings are of long duration and may reach ranges of the order of 10,000 km., the transmitter power must be much higher than that of an ordinary radio-sonde, thus necessitating higher capacity batteries and greater weight. Further, since the transmission may not take place in the line of sight the use of very high frequencies (30 Mc./s. and above) is not advisable. It is, in fact, desirable to select frequencies that are suitable for the particular propagation conditions of the time and season of use; generally, specific frequencies within the band 1 to 20 Mc./s. are suitable. Another important consideration arises if the measurements during level flight include temperature and humidity, for since there is no relative motion between the balloon and the air such ventilation of the measuring elements as is provided by the swinging of the radio-sonde on its suspension may be inadequate and some means of artificial ventilation may be necessary.

The application of radio-sondes to long-range balloon soundings in the U.S.A. has been described by Spilhaus and his associates[92] and, more recently, by Anderson and Mastenbrook[93]. Standard American designs of radio-sonde were adapted for this purpose. The baroswitch type of pressure indicator was replaced by a motor-driven helical scanner or coded disc scanner so as to permit the determination of pressure independently of a knowledge of the contact sequence which is essential with the baroswitch. A conventional transmitter, with an output power of 50 W., was used. It was switched on by a d.c. timing motor every two hours for five minutes on each of three frequencies of about 6, 12 and 19 Mc./s. The meteorological signals were interrupted during each transmission by a signal of 30 sec. duration for radio direction-finding. On some occasions trajectories as long as 10,000 km. have been tracked with this equipment.

British development of a long-range radio-sonde has, as far as is known, been confined to an experimental instrument designed in the Radio Division of the National Physical Laboratory about 1945. With a power output of 7 W. from a crystal-controlled oscillator using frequencies of 1·7, 2 or 3 Mc./s. according to the season, this instrument was intended to be used for ranges up to about 1500 km.

and flight durations of 12 hr. Transmission was made for one minute every fifteen minutes during level flight, the signal being used for direction-finding, and a modulation frequency to determine pressure. At a pre-determined time after launching, the radio-sonde was released automatically from the balloon and, during its descent by parachute, signals were emitted continuously with modulation frequencies indicating pressure, temperature and humidity.

7.3.2. Drop-sondes

The launching of radio-sondes from aircraft affords another way of obtaining upper air soundings in areas where launching from the ground is impracticable. It has the advantage over the long-range balloon method in that it allows the time and place of the descent to be arranged more precisely, but it may involve uncertainty about the height of the final pressure signal. In order to survive the process of ejection from an aircraft the drop-sonde has to be more robust than the balloon-borne instrument. Specially designed release gear is required for launching the sonde and its parachute without disturbing the calibration. If the aircraft is pressurized the release gear must have an air-lock system and the radio-sonde must be fitted externally. In the latter case means must be provided for keeping the radio-sonde battery from becoming too cold for efficient use.

There are two possible arrangements for the transmission link between the drop-sonde and the base station, namely, direct or via the aircraft. For direct transmission over long ranges to receiving equipment on the ground it is necessary, as in the case of constant-level balloon flights, to use a transmitter of relatively high power (of some tens of watts) working on radio-frequencies in the H.F. band (3 to 30 Mc./s.). The advantage of using the aircraft as an intermediate link between the drop-sonde and the ground station is that transmission of relatively low power on V.H.F. (30 to 300 Mc./s.) can be used between the sonde and the aircraft. It is then relayed to the ground through the aircraft H.F. communication channel. This system necessitates the aircraft remaining within V.H.F. range, e.g. 300 km., while the sonde is descending and also being within communication range of the ground station. Since the aircraft may be subject to a higher level of interference than a ground station the power of the sonde transmitter should be somewhat higher than that of the normal radio-sonde, but it may be considerably less than is required for direct H.F. transmission. An alternative indirect arrangement using the V.H.F. sonde-to-aircraft link is to install complete radio-sonde receiving or recording equipment in the aircraft. This has a doubtful advantage over the relay system in that it permits of operations beyond the range of the aircraft communication system, though at the cost of delaying the receipt of the information at the ground station. Clearly, a radio-sonde system involving heavy or bulky equipment would not be acceptable for this arrangement and a system such as the code-type, for which an ordinary radio receiver suffices, is the most suitable.

A code-type of instrument which has been used as a drop-sonde in the U.S.A. has been described by Brailsford[94]. The transmitter is a single-valve, continuous-wave, crystal-controlled oscillator operating on a specific frequency in the range 2 to 12 Mc./s. with a power of 0·3 W. The oscillations are interrupted by a keying relay operated by " pick-up " arms which are actuated by the meteorological units; these units comprise a bimetal thermometer, aneroid barometer and hair hygrometer. Coded signals are produced by contacts of the arms and the grooves of a rotating sector, of angle 85°, forming a raised surface on a disc. There are 200 grooves carrying the various code groups, each groove corresponding to steps of

5 mb. in pressure, 0·8°C. in temperature and 2 per cent. relative humidity. Except when the pick-up arms are on the sector they are quite free to respond to the action of the sensitive elements. A small, self-starting, uni-directional electric motor drives the disc at 12 revolutions per minute, thus providing one cycle of readings for every 30 m. of height in a descent at 6 m./sec. The bimetal thermometer is made in the form of an elliptical spring, in which the inner leaves are of the metal with the higher expansion coefficient. This design was selected as being able to withstand the high acceleration at launching (which may reach 20*g*) and also because it permits the use of thin material, thereby giving rapid response. It was also found to have the advantage of a linear temperature characteristic. The apparatus incorporated a parachute specially designed to minimize the shock at launching. It has been launched from aircraft at heights up to 10 km. at speeds of 260 kt., the signals being received in the aircraft on an ordinary receiver covering the frequency in use.

7.3.3. Applications to atmospheric electrical measurements

There have been a number of applications of radio-sonde technique to the measurement of atmospheric potential gradient and electrical conductivity and a few of them will be briefly described. One of the earliest is due to Kreielsheimer and Belin[95] who adapted the principle of the alti-electrograph (see Section 1.3.1) to radio-sonde telemetering. In the alti-electrograph the variation with height of the sign of the potential gradient during thunderstorms and showers was recorded in the balloon-borne instrument by detecting the point-discharge current which flows through a vertical wire when the potential gradient exceeds the value required to start corona discharge at the ends of the wire. Belin's adaptation gave quantitative measurement as well as the sign of the potential gradient and transmitted the measurements by radio link. He used the earlier, Diamond–Hinman, type of American radio-sonde in which the radio-frequency oscillator, transmitting on a frequency of about 70 Mc./s., is modulated by a blocking oscillator, the grid potential of which is controlled by resistive type meteorological units (see Section 7.2.4). In Belin's modification the point-discharge wire, 20 m. long, has at its mid-point a resistor in the grid circuit. The current through the wire therefore controls the potential on the grid and the modulation frequency so produced is recorded by the ground equipment in the normal manner. Chapman[96] used a similar method with a later type of American radio-sonde operating on 403 Mc./s. The modification in this case involved the addition of a valve, on the grid resistor of which the point-discharge current produced a voltage drop, which in turn controlled the bias of the blocking oscillator.

The Meteorological Office radio-sonde (described in detail in Chapter 8) has been adapted for atmospheric electrical measurements by Koenigsfeld and Piraux[97]. In their system, which is primarily intended for use in non-stormy weather when the potential gradient is too small to cause point discharge, two polonium collectors separated vertically by a distance of one or two metres are used instead of the long wire. The potential difference between the collectors is applied to the anode and cathode of a triode electrometer valve, from the grid of which current is fed into the coil of an inductor similar to that used in the meteorological unit but with a fixed armature gap. Variations of the potential gradient in turn vary the grid current which controls the degree of polarization of the fixed inductance, thus giving variations in inductance, and hence frequency, which are transmitted in the usual way. This circuit requires the anode of the electrometer valve to be negative.

A simplified diagram of the arrangement is shown in Fig. 20. A similar scheme was developed by Koenigsfeld[98] for measuring the electrical conductivity of the air. In this case the plates of the cylindrical condenser of a standard type (Gerdien) of conductivity apparatus are connected to the anode and cathode of the electrometer valve; the potential difference then depends on the rate of discharge of the condenser by atmospheric ions.

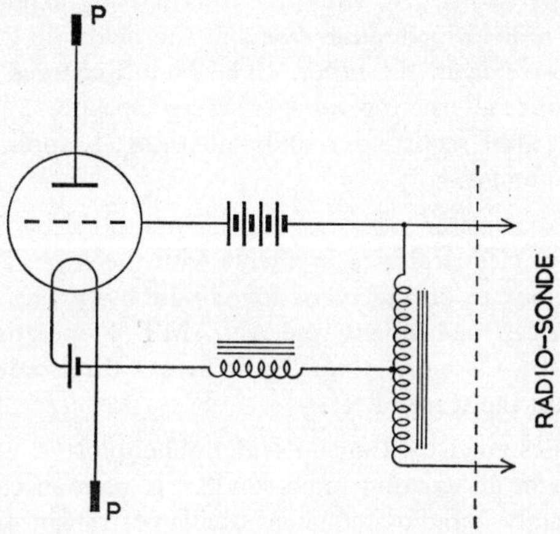

FIG. 20—KOENIGSFELD'S ADAPTATION OF THE METEOROLOGICAL OFFICE RADIO-SONDE FOR POTENTIAL GRADIENT MEASUREMENT
P, polonium collectors.

A more recent application of the Meteorological Office radio-sonde to the measurement of vertical electric field and polar conductivity has been described by Jones, Maddever and Sanders[99]. Its chief feature is the use of a current-regulated inductance for controlling the audio-frequency signal. This arrangement, which is more sensitive than Koenigsfeld's method, is illustrated in Fig. 21. A mumetal core carries the windings of the oscillator coil in two halves A and B, the control coil C and over this a separate coil D of much fewer turns. When A and B are plugged into the radio-sonde in place of the standard inductor any currents through either of the other two windings increases the oscillation frequency. The d.c. output from an electric field or conductivity measuring device is passed through C while D is connected through a potentiometer to a $1 \cdot 5$ V. cell to provide

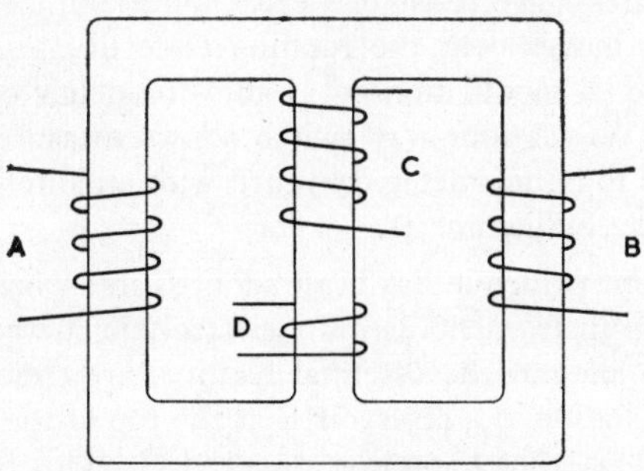

FIG. 21—CURRENT-REGULATED INDUCTANCE USED BY JONES, MADDEVER AND SANDERS[99]

a setting adjustment. The normal range of audio-frequency variation from 700–1000 c./s. can be obtained with as little as 13 μ amp. The conductivity measurement is made with a Gerdien type of apparatus and the ion currents are amplified by an electrometer valve circuit before passing through the control coil of the inductor. Instead of employing the radioactive collector or point-discharge method for the electric field measurement, a double " field mill " is used. This consists essentially of two plates which are alternatively exposed to and shielded from the atmospheric field in the one case and the field due to any charge on the surface of the radio-sonde in the other. The bound charges are proportional to these fields and produce alternating voltages across capacitors. These are amplified and their difference, after rectification and smoothing, is applied as a signal to the control coil of the inductor.

Koenigsfeld's electrometer valve has also been applied by Venkiteshwaran and his associates[100, 101] to electrical measurements with the American 70 Mc./s. radio-sonde. A more elaborate arrangement for conductivity measurement using the more modern American instrument, the AN/AMT–4 operating on 1680 Mc./s., has been described by Coroniti *et alii*[102]; it uses a three-valve d.c. electrometer amplifier to modulate the transmitter.

An instrument designed by Lugeon and Bohnenblust[103] is noteworthy in not being an adaptation of an existing radio-sonde. It uses an electrometer valve to pulse-modulate directly a radio-frequency oscillator transmitting on a frequency of 400 Mc./s. As no meteorological quantities other than potential gradient or conductivity are measured by the instrument it must be sent up in tandem with an ordinary radio-sonde for heights to be determined.

7.3.4. Applications to ozone measurements

Coblentz and Stair[104] in 1938 determined the vertical distribution of ozone by measuring the ultra-violet intensity of solar radiation with a cadmium photocell placed in the grid circuit of the blocking oscillator of a Diamond–Hinman radio-sonde. A series of filters was rotated over the photocell by an electric motor so that the intensities of several narrow bands of the solar spectrum could be investigated.

The same principle has been adopted by Kulcke and Paetzold[105] for use with a German code-type radio-sonde. A selenium photocell with a quartz window is used and the incident light, after passing through a blue or ultra-violet filter, is modulated at a frequency of 50 c./s. by a rotating shutter. The output of the photo-cell goes through a three-stage resistance–capacitance amplifier to a milliammeter, the pointer of which moves over the rotating code disc. The code signals are transmitted as a 600 c./s. modulation on a radio-frequency of 152 Mc./s. with an output power of 0·2 W. A four-way switch allows measurements with the blue and ultra-violet filters to be telemetered in turn with an amplifier reference voltage and a zero check of the milliammeter.

An entirely different principle has been adopted in an ozone sonde devised by A. W. Brewer. This instrument uses an electrolytic method of measuring the ozone content of the air and the essential features are shown in Fig. 22(*a*). A solution of potassium iodide in a reservoir R at the top of the apparatus flows very slowly through a capillary tube C on to a glass rod G. This rod is wound for most of its length with a platinum wire spiral to act as cathode and with a single turn of wire just below this to act as anode. By means of a miniature rocking piston pump driven by a toy electric motor air is drawn through the entrance P down the gap A between the glass rod and the cylinder encasing it. The whole unit is thermally insulated to prevent the solution from freezing.

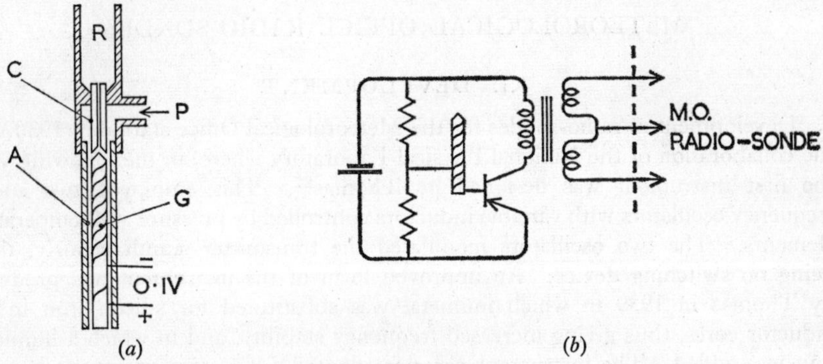

FIG. 22—OZONE SONDE DEVISED BY A. W. BREWER

(*a*) Ozone measuring apparatus.

(*b*) Transistor amplifier and telemetering circuit.

With a potential difference of 0·1 V. across the electrodes no current flows so long as the potassium iodide solution remains pure. If, however, the air contains ozone then iodine is formed in the solution by chemical oxidation, thus:

$$2KI + O_3 + H_2O = 2KOH + O_2 + I_2 .$$

This iodine is reduced electrolytically at the cathode, thus:

$$I_2 + 2e \rightarrow 2I' ,$$

and conducts current to the anode. There the reverse action takes place and the iodine liberated is washed away by the stream of solution. The strength of the current is, therefore, determined by the rate of absorption of ozone. With an air flow of 2000 cm.³/min., a current of a few microamperes may be obtained. This is converted into a radio-sonde signal by first amplifying it with a single transistor, as shown in Fig. 22(*b*), and then passing the amplified current through the central coil of a mumetal inductor of a Meteorological Office radio-sonde, thereby producing an audio-frequency modulation as in the normal use of the instrument.*

* Since the above description was written full details of the ozone sonde, including an alternative design of electrolytic detector, have been published by Brewer and Milford[142].

CHAPTER 8

METEOROLOGICAL OFFICE RADIO-SONDES

8.1. DEVELOPMENT

Development of radio-sondes for the Meteorological Office started in 1936 with the collaboration of the National Physical Laboratory where, in the following year, the first instrument was designed by Thomas[27]. This employed two audio-frequency oscillators with variable inductors controlled by pressure and temperature elements. The two oscillators modulated the transmitter simultaneously, there being no switching device. An improved form of the instrument was produced by Thomas in 1939 in which mumetal was substituted for silicon iron in the inductor cores, thus giving increased frequency stability, and in which a humidity unit was added. The instrument was not adopted for routine use partly because it was very heavy (2900 gm. with its battery) and partly because the temperature and humidity units were not entirely satisfactory in operation. A Mark 3 version of the N.P.L. design from which the humidity unit was omitted was, however, produced and used in small numbers in 1940.

A new design, again based on the variable inductance principle, was developed by Dymond[28] with the assistance of other members of the Meteorological Office at Kew Observatory in 1940. The new instrument became known as the Kew radio-sonde (Mark 1). It differed from the N.P.L. design in having only one audio-oscillator, to which each inductor was connected in turn by means of a windmill-driven switch. This allowed a reduction in weight to about half that of the N.P.L. design but it involved the sacrifice of two advantages of that design, namely, the facility for continuous measurement and the total absence of moving parts other than those of the meteorological elements. The latter also were fundamentally different in the two designs, the Kew instrument using a bimetal strip instead of a brass wire in an invar frame for temperature measurement, a single aneroid capsule instead of metal bellows for pressure, and gold-beater's skin in place of hair for humidity. Adoption of a wet acid battery in place of dry cells accounted for a further saving of weight of about 600 gm. A modified version of the Kew Mark 1 instrument in which the temperature and humidity units and the windmill switch were omitted was used for upper wind measurement by the direction-finding method. This instrument was known as the W.F.6 transmitter.

In 1943 a new model of the Kew radio-sonde, the Mark 1A, was introduced. This differed from its predecessor mainly in the substitution of grid modulation for plate modulation of the radio-frequency valve, thus eliminating the audio-frequency choke. Other changes were the replacement of the flat strip form of the bimetal element by one of cylindrical shape, and the improvement of the radiation shield.

In 1945 the Kew radio-sonde was redesigned for more economical manufacture by mass-production methods of tooling. The new model was known as the Meteorological Office radio-sonde Mark 2, but it did not differ in principle from the earlier models. The design has, in fact, continued practically unchanged up to the present model, the Mark 2B, which was introduced in 1952, the main changes then being the adoption of a quarter-wave instead of a half-wave end-fed aerial, the provision of temperature compensation in the radio-frequency circuit, the reduction of temperature and voltage variations on the audio-frequency stage, and improvement in the performance of the inductors.

A requirement for a radio-sonde of high precision led to development being started in 1944 of an instrument on the chronometric principle for use with a microwave transmitter which could be tracked for wind-finding by a radar theodolite (see Chapter 5). The scanning unit was in the form of a helix on a cylinder driven by an electric motor. Absolute synchronism between the rotation of the helix and the ground recorder was ensured by synchronizing pulses generated at a rate of about 33 per second. A prototype instrument was completed in 1946 but it was concluded that the high precision required would be difficult and expensive to achieve in production models. By 1948 it had been decided to adopt pulse technique in the airborne transponder for use with a radar theodolite and this presented the attractive possibility of using the microwave pulse transmission for the additional purpose of telemetering the information from meteorological elements. The research on such a radar-sonde system was undertaken by the Ministry of Supply and was described in 1951 by Jones and his associates[37]. The subsequent development was undertaken in the Mullard Research Laboratory.

To meet the need for an instrument for signalling temperature only, and light enough to be carried up to 6 km. on a 100 gm. balloon, a variable audio-frequency radio-sonde using a fixed inductor and a variable capacitor was developed in 1953 by R. Almond. The dielectric of the capacitor acts as the temperature-sensitive element in this light-weight radio-sonde.

8.2. THE METEOROLOGICAL OFFICE RADIO-SONDE, MARK 2B

A detailed account of the Kew radio-sonde and its performance has been given by Dymond[28], and much of the information given in that account applies equally well to the Mark 2B instrument which is of the same basic design. A schematic diagram of the radio-sonde is given in Fig. 23 and a photograph of the instrument is reproduced in Plate XXV.

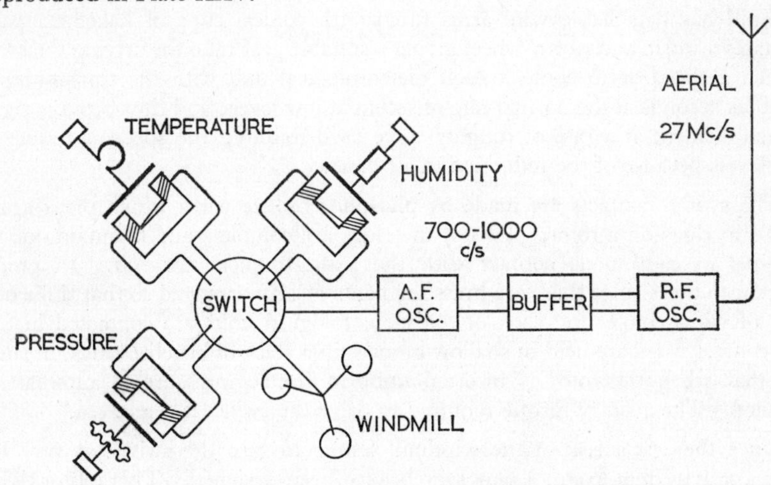

FIG. 23—SCHEMATIC DIAGRAM OF METEOROLOGICAL OFFICE RADIO-SONDE, MARK 2B

8.2.1. Principle

The transmitter of the Mark 2B radio-sonde is designed to operate within the frequency band 27·0 to 28·5 Mc./s. The carrier wave is modulated through a buffer circuit by an audio-frequency oscillator, the frequency of which is controlled by the three meteorological units measuring pressure, temperature and humidity. A buffer stage is used in order to prevent reaction between the radio-frequency and audio-frequency circuits. The oscillator-tuned circuit has a fixed capacitor and is connected, by means of a double-pole three-way switch, in turn to three variable inductors operated by the meteorological elements. The three inductors each consist of two coils, with a mumetal core and a movable mumetal armature, the position of which is controlled by an aneroid capsule, a bimetal element and a strip of gold-beater's skin respectively. As the armature approaches the core the air gap in the magnetic circuit is reduced, the inductance rises and the frequency of the oscillating circuit falls. The audio-frequency variation is within the range 700 to 1000 c./s. Electrical power is provided by a special light-weight battery, while the power to drive the switch is provided by a three-cup windmill but this may be replaced in the near future by a gravity motor (see Section 8.2.2).

Reception of the signals is made with a sensitive short-wave receiver, the pitch of the audible modulations being measured by matching it with the note produced by a calibrated oscillator of the resistance–capacity type, the matching being done on a cathode ray oscilloscope by obtaining a one-to-one Lissajous pattern.

8.2.2. Construction

The transmitter and the battery are enclosed in a pressed-cardboard cylindrical case made rigid and waterproof by resin bonding and consisting of two parts which fit above and below a moulded bakelite circular panel (or chassis) on which the transmitter is mounted, with the switch and most of the radio components underneath and the three valves and other components on top. Protruding from the case are the three meteorological units and the windmill. The meteorological units plug into sockets in the panel; when viewed from above the units are arranged in clockwise order, thus: windmill, hygrometer, barometer, thermometer. The windmill has detachable wire arms fitted with conical cups of bakelized paper. It rotates a worm and worm-wheel giving a suitable gear ratio for driving the switch at such a speed as to connect each meteorological unit with the transmitter for about six seconds at the normal rate of ascent at low levels, and thus provide signals of each element at a rate of roughly three each minute; the speed is reduced at high levels because of the reduction in air density.

The switch contacts are made by phosphor bronze wires which press against brass slip rings on a rotor. Two of the rings are complete and maintain one pair of wires in continuous contact with the audio-frequency circuit. In grooves adjacent to these rings there are brass segments of 120° arranged so that three other pairs of wires connecting each of the meteorological units are contacted in turn. The contact wires are held in shallow grooves in a stack of bakelite rings, in such a way that when the rotor is inserted uniform contact pressure is automatically obtained. The average torque required to drive the switch is 4 gm. cm.

Since there is a risk of the windmill failing to turn the switch at very high levels a gravity motor and a new switch have been designed. This motor, which is fitted on the upper part of the transmitter panel, consists of an arrangement of gear-connected spindles. One of these carries a drum wound with suspension cord and forms the prime mover. A second spindle operates the switch, and the final spindle carries a speed-governor of centrifugal friction type that is independent of air resistance. The motor is operated by the load on the suspension cord, i.e. the weight of the radio-sonde. In the switch the slip rings and arcs are in printed circuit form. The sequence of contacts is such that in each cycle the signals are in the order: pressure, temperature, humidity, temperature.

The weight of the radio-sonde, complete with meteorological units, windmill, case and aerial is 885 gm. The battery, charged with acid and lagged with wadding, together with its lead and plug weighs 350 gm., making a total weight of 1235 gm.

8.2.3. Transmitter

A diagram of the electrical circuit is reproduced in Fig. 24. All three valves are triodes of type CV 1586, taking 55 m. amp. filament current at 2·3 V. The audio-frequency generator consists of a Hartley oscillator V1 with a fixed capacitor C1 of 0·07 μF. capacitance. The value of this capacitance, in combination with the variable inductances controlled by the meteorological elements, determines the oscillation frequency. C1 must therefore possess high stability and have a low temperature coefficient. A wax-coated, clamped, silvered mica capacitor, with a temperature coefficient of less than 35×10^{-6} per °C., is normally used. In order to stabilize the oscillation frequency against change in battery voltage the capacitance–resistance combination C2R2 is inserted between the oscillating circuit and the driving valve. The stabilization is such that the overall variation due to the average drop in battery voltage during an ascent rarely exceeds 0·1 c./s. The temperature coefficient of the audio-frequency circuit is such that if operated with

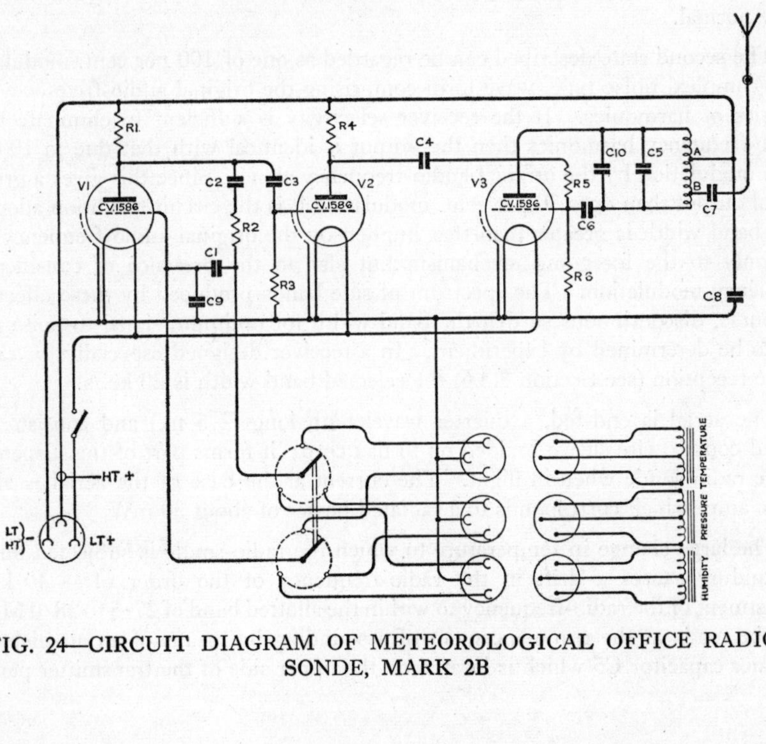

FIG. 24—CIRCUIT DIAGRAM OF METEOROLOGICAL OFFICE RADIO-SONDE, MARK 2B

a fixed inductor held at a fixed temperature the frequency does not change by more than $\pm$ 0·5 c./s. for the full temperature range experienced in flight. This is very important since only the inductors are subjected to the normal calibration cycles of pressure, temperature and humidity. Another important component is the capacitor C9 which serves to decouple the grid of V1 from radio-frequency oscillations arising in the transmitter and so make the variations due to this effect less than 0·2 c./s.

The oscillations generated in the audio-circuit are applied to the grid of the valve V2 through a 0·001 μF. capacitor and across a 50,000 Ω grid leak, thus imposing a constant load on the audio-oscillator and maintaining its stability. The feed capacitor must be sufficiently large to cause effective modulation of the radio-frequency. Some amplification occurs in this buffer stage but its principal function is to isolate the high-frequency and low-frequency oscillators and to prevent interaction between them.

The radio-frequency circuit which is temperature-compensated is of the conventional Hartley type, with grid modulation of the single valve V3 by the output of V2. The depth of modulation is controlled by the coupling resistor and capacitor R5 and C4. It is found that with this arrangement two states of modulation are possible. In one the modulation depth is 30 per cent. or less and the audio-frequency wave form has little distortion. In the other over-modulation occurs, both peaks of the modulating wave form being heavily clipped and the radio-frequency output comprising a series of rounded pulses at the modulating frequency with a " mark/space " ratio of about 1 to 3. Depths of modulation between these two states cannot be achieved. The signal strength, as measured by a receiver, depends not only on the aerial current, i.e. the amplitude of the carrier wave, but also on the depth of modulation. Maximum signals are received if modulation is 100 per cent., i.e. if the carrier amplitude varies between zero and twice normal.

The second state described can be regarded as one of 100 per cent. modulation by a complex, pulse type, wave form comprising the original audio-frequency and a series of harmonics. If the receiver selectivity is sufficient to eliminate these audio-frequency harmonics then the output is identical with that due to 100 per cent. modulation by the original audio-frequency alone. Since this gives a greater useful output than does 30 per cent. modulation it is the circuit condition adopted. The band width is greater than that implied by the original audio-frequency due not only to the foregoing mechanism but also to the presence of considerable frequency modulation. The spectrum of side bands produced by these effects is, of course, discontinuous so that the band width for optimum signal-to-noise ratio has to be determined by experiment. In a receiver designed especially for radio-sonde reception (see Section 8.2.6) the selected band width is 20 kc./s.

The aerial is end-fed, a quarter wavelength long (2·5 m.) and consists of a tinned copper wire sleeve braided on to flax cord; it forms part of the suspension of the radio-sonde when in flight. The current at the base of the aerial is about 30 m. amp. which corresponds to a radiated power of about 30 mW.

The large change in temperature to which the radio-sonde is subjected during a sounding causes a drift in the radio-frequency of the order of + 40 kc./s. Adjustment of the radio-frequency to within the allotted band of 27·5 to 28·0 Mc./s. may be made at the commencement of a sounding by means of an air-dielectric trimmer capacitor C5 which is located on the upper side of the transmitter panel.

8.2.4. Meteorological units

These units, each comprising a meteorological element and inductor protected by a radiation shield, are detachable from the transmitter panel for the purposes of calibration and packing. In use they are exposed to the air. This is of importance since the inductors are sensitive to temperature and in order to apply corrections their temperature must be known.

The inductors.—These are all alike and consist of a U-shaped laminated core of mumetal stampings round which two coils, each of 1250 turns of wire, are fitted, and a mumetal armature mounted on a nickel-silver spring hinge. Mumetal has a very high permeability in weak magnetic fields and in order to obtain the maximum value the stampings and the armature are heat-treated. They are, however, exceedingly sensitive to rough treatment and the units must, therefore, always be handled with care. The permeability also changes with temperature, a fact which is of particular importance in the pressure unit.

The two coils are in series, with their junction earthed and the other ends connected through the switch across the 0·07 μF. capacitor, thus forming the " tank " or tuned circuit of the audio-oscillator. Each armature is attached to a meteorological element, which thus controls the air gap between the armature and the pole pieces of the inductor core and so varies the tuning of the oscillator. The useful range of movement of the armature is about 2·5 mm. and the sensitivity in terms of the change in audio-frequency varies from about 50 c./s. per millimetre at 1000 c./s., when the gap is wide, to 200 c./s. per millimetre at 700 c./s. This non-linearity has the important advantage of increasing the sensitivity of the pressure unit by a factor of about three at great heights. It also compensates for a falling off in sensitivity of the bimetal temperature element at low temperatures. To achieve these advantages the inductors are designed so that high audio-frequencies correspond with surface values of pressure and temperature.

The pressure element.—This is an evacuated steel aneroid capsule. It gives an approximately linear relation between deflexion and pressure, the deflexion being about 2 mm. for a change of pressure of 1000 mb. The corresponding frequency change is, therefore, about 200 c./s.

Elastic hysteresis shows itself as a slight difference in calibration when a capsule is submitted to pressure changes in opposite directions. This has the effect that a significant point, such as a sharp inversion, on the graph of a sounding appears to lie at a slightly higher level on the descent than on the ascent. The Meteorological Office specification stipulates that the difference of pressure at 500 mb. when decreasing and increasing over the cycle 1000 to 50 to 1000 mb. should not exceed 2 mb. Units are all calibrated with a falling pressure so that calibration can be considered accurate on the ascent.

The pressure unit as a whole is sensitive to temperature. Most of the effect is due to the mumetal of the inductor, which is subject to variation in magnetic hysteresis and eddy current, and also in permeability, with temperature. Resistance of the coils changes by about 30 per cent. between + 15° and — 60°C. and causes a change of audio-frequency of about 4 c./s. The aneroid capsule itself has a temperature coefficient of its elastic constant. This effect is greatest at ground level, where the stress on the capsule is a maximum; it is, however, small compared with the effects of the mumetal and the coils and no attempt has been made to compensate for it. The total temperature effect in the pressure unit is not a linear function of either temperature or frequency and it also varies widely from one instrument to another. This variation is now reduced by the individual testing and grading of inductor laminations. Average values are shown in Fig. 25 where the frequency change is plotted against oscillator frequency for various temperatures. It will be seen that at the lower frequencies, corresponding to lower pressures, the differences are quite large and may reach 20 c./s. The procedure for applying the necessary corrections is described in the *Handbook of radio-sounding technique*[46].

The temperature element.—This is a bimetallic strip, 0·025 cm. thick, rolled into a cylinder 1 cm. diameter by 1·6 cm. high, which curls up with increase of temperature. A combination of ordinary and invar steel is used and after assembly the element is seasoned by first maintaining it at about 100°C. for 8 hr. and then by taking it through a temperature cycle from + 60° to − 75°C. and back to + 60°C. at least five times. This relieves internal strain and improves stability. At low temperatures the sensitivity of the bimetal element is reduced owing to the increase in Young's modulus and the design of the linkage, but this is offset by the increasing sensitivity of the inductor for small gaps and the combination gives a smooth variation of sensitivity from 0·7° to 0·4°C. per c./s. over a range from + 45° to − 85°C. The temperature coefficient of the inductor itself is the same as that in the pressure unit but this does not necessitate special correction since it is included in the calibration.

The speed of response of the temperature element in its radiation shield, expressed as the time required for an initial temperature difference between the element and the air to be reduced in the ratio $1/e$, is 7·4 sec. at ground level in an air flow of 6 m./sec. (about the normal rate of ascent). The lag increases rapidly at high levels, becoming 97 sec. at a pressure of 10 mb. At this pressure level the lagging effect of the shield, which is almost negligible at ground level, is quite large; without its shield the bimetal element has a lag coefficient of 56 sec. at 10 mb.

The radiation shield is in the form of a duct of rectangular cross-section and is made of thin polished aluminium. The lower half surrounding the element is double-walled and the upper half is an extension of the inner wall. This arrangement is reasonably effective in the lower levels but at high levels, where the low air density reduces the cooling effect of forced convection, the air reaching the element is heated appreciably in day-time by the shield itself. A detailed study of the radiation and lag errors of the Meteorological Office radio-sonde and the evaluation of temperature corrections has been made by Scrase[106, 107]. The procedure for the application of corrections to routine measurements is given in the *Handbook of radio-sounding technique*[46].

The humidity element.—This element consists of a strip of single-ply unvarnished gold-beater's skin 6·4 mm. wide with an effective length of 15 mm. when mounted in the humidity unit, in which it is under a tension not exceeding 20 gm. when the inductor armature is 3 mm. from the pole pieces. With increase in humidity the skin increases in length and the tension, therefore, decreases. After mounting in the unit the element is seasoned for half an hour in a saturated atmosphere circulating at 2 m./sec.; this treatment is essential if reproducible results are to be obtained. As with hair, the skin measures relative humidity with respect to liquid water, irrespective of the temperature.

Gold-beater's skin exhibits a hysteresis effect when subjected to a cycle of humidity change which includes dry conditions (less than 30 per cent.), but it recovers its calibration when it returns to above 70 per cent. relative humidity. It is advisable, therefore, to condition the element by placing it in a saturated

SECTIONAL VIEW OF METEOROLOGICAL OFFICE RADIO-SONDE, MARK 2B

Plate XXVI

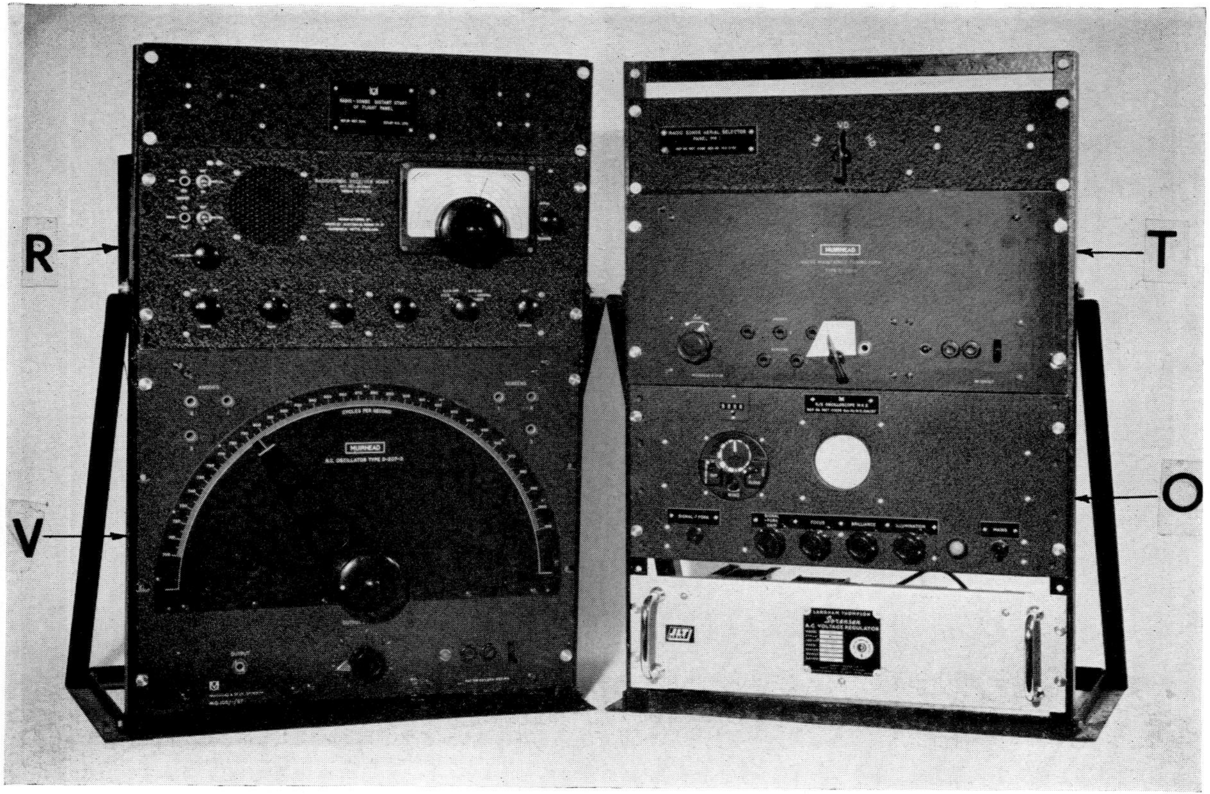

GROUND EQUIPMENT USED WITH METEOROLOGICAL OFFICE RADIO-SONDE, MARK 2B
R, receiver; V, variable audio-frequency oscillator; T, valve-maintained tuning fork; O, oscilloscope.

GENERAL VIEW OF AUTOMATIC RECORDING EQUIPMENT FOR METEOROLOGICAL OFFICE RADIO-SONDE, MARK 2B

Plate XXVIII

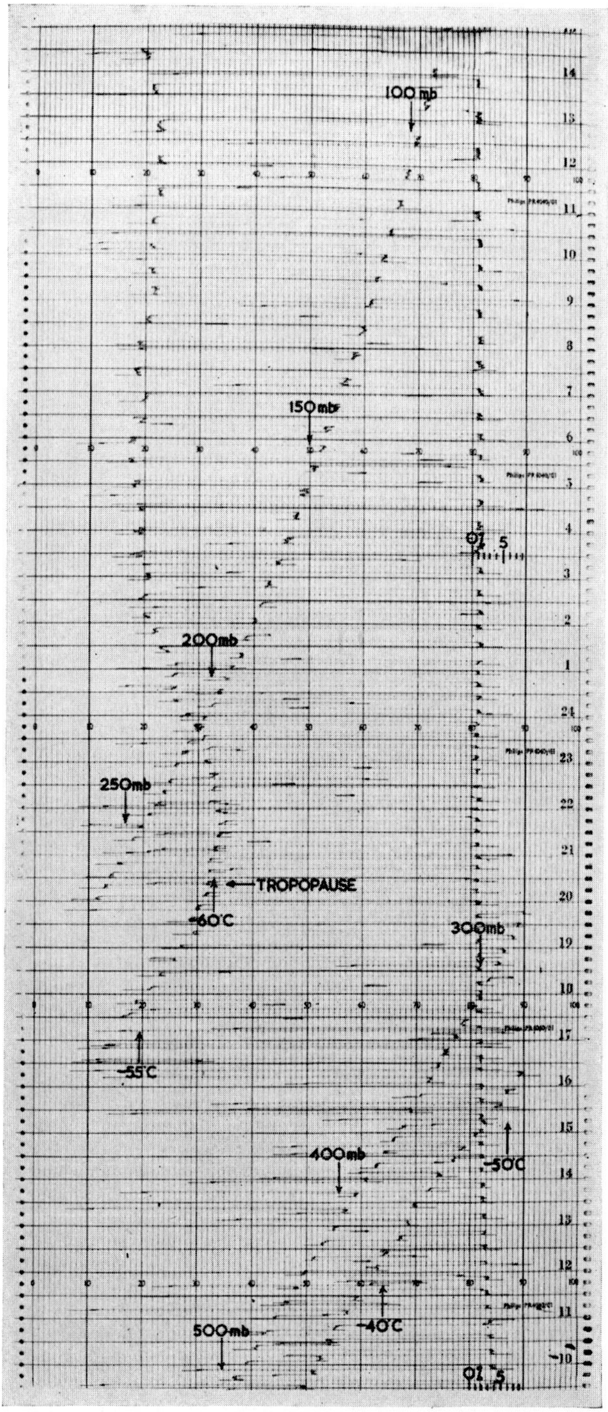

AUTOMATIC RECORD OF METEOROLOGICAL
OFFICE RADIO-SONDE SIGNALS

The left-hand edge of the recorded deflexions represents the value measured.

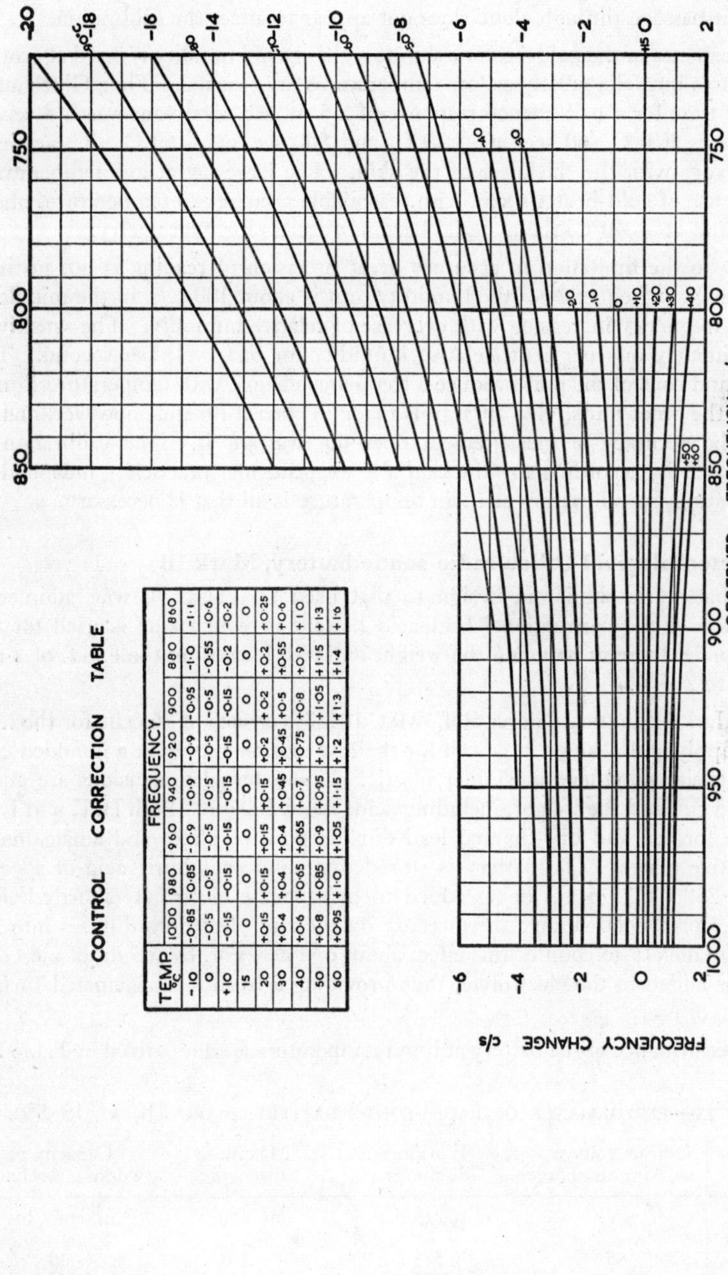

CONTROL CORRECTION TABLE

TEMP. °C	FREQUENCY 1000	980	960	940	920	900	880	860
-10	-0·85	-0·85	-0·9	-0·9	-0·9	-0·95	-1·0	-1·1
0	-0·5	-0·5	-0·5	-0·5	-0·5	-0·5	-0·55	-0·6
+10	-0·15	-0·15	-0·15	-0·15	-0·15	-0·15	-0·2	-0·2
15	0	0	0	0	0	0	0	0
20	+0·15	+0·15	+0·15	+0·15	+0·2	+0·2	+0·2	+0·25
30	+0·4	+0·4	+0·4	+0·45	+0·45	+0·5	+0·55	+0·6
40	+0·6	+0·65	+0·65	+0·7	+0·75	+0·8	+0·9	+1·0
50	+0·8	+0·85	+0·9	+0·95	+1·0	+1·05	+1·15	+1·3
60	+0·95	+1·0	+1·05	+1·15	+1·2	+1·3	+1·4	+1·6

FIG. 25—PRESSURE CORRECTION GRAPH FOR METEOROLOGICAL OFFICE RADIO-SONDE, MARK 2B

atmosphere for at least 20 min., and preferably 60 min., both before calibration and use, and to calibrate from damp to dry conditions (the overall direction of humidity change during a sounding). This treatment does not prevent further permanent change in length if the gold-beater's skin becomes wet with liquid water, so in order to protect it from rain a metal shield is fitted above it in the unit. The shield is not effective against prolonged exposure to extremely wet fog during preparation for an ascent, but passage through cloud does not appear to affect the calibration.

The response of the gold-beater's skin to a change of humidity is rapid at ground temperatures but falls greatly as low temperatures are reached. Thus Glückauf[108] found the time for a half-change response in a 5 m./sec. airstream was 2·4 sec. at 18°C., 6 sec. at 0°C., 60 sec. at − 30°C. and 800 sec. at − 60°C. These times probably vary with the thickness of the skin. The large lag at low temperatures limits the use of gold-beater's skin with reasonable accuracy to temperatures above − 20°C. and at − 40°C. the material becomes useless for hygrometry.

Owing to the limitation in accuracy great precision of reading is not justified and the range of frequency of the humidity unit is about 100 c./s. in the middle of the band, frequency increasing with decrease of relative humidity. The sensitivity is approximately one per cent. relative humidity for one cycle per second. The mumetal and coils of the unit produce a frequency change with temperature similar to that of the other units. No attempt is made to correct for this, however, mainly because the accuracy of measurement does not warrant it. The calibration of the skin itself is, according to Glückauf's[108] experiments, practically independent of temperature, so calibration at room temperature is all that is necessary.

8.2.5. Meteorological Office radio-sonde battery, Mark 1B

This battery is similar in design to that used by Väisälä[79], who adopted a modification of the conventional lead–acid battery in which zinc is used for the negative plate, thereby reducing the weight and also increasing the e.m.f. of a cell from 2·0 to 2·4 V.

The Mark 1B battery (Stores Ref., Met. 10453) consists of 36 cells for the high tension supply and a single large cell for the low tension supply in a moulded case of high density polythene or similar plastic. The external dimensions are about 7 × 7 × 5 cm. and the weight, including acid, about 300 gm. Both H.T. and L.T. cells have formed and dry-charged lead dioxide positive plates and amalgamated zinc negative plates. The battery is stored dry, dilute sulphuric acid of specific gravity 1·250 (at 15·5°C.) being added to provide the electrolyte shortly before use. After prolonged storage the mercury on the zinc electrodes diffuses into the body of the metal; to counter this effect about one per cent. by weight of mercuric sulphate is added to the electrolyte, thus providing a freshly amalgamated surface at the time of use.

The performance of the battery at room temperature is summarized in Table IX.

TABLE IX—PERFORMANCE OF RADIO-SONDE BATTERY, MARK 1B, AT 15·5°C.

Section	Output voltage at working discharge	Working discharge	Maximum discharge	Capacity at working discharge
	V.	m. amp.	m. amp.	m. amp. hr.
H.T.	85	5·5	10	11
L.T.	2·3	175	250	350

At the working discharge rates the output voltages at room temperature remain constant to within about 2 per cent. for 2 hr. At low temperatures the efficiency of the battery depends on the rate of discharge; the e.m.f. changes little, but the internal resistance rises with fall of temperature. With the normal rate of discharge the cells begin to fail at temperatures about −30°C. and with the maximum rate they begin to fail at about −15°C. In practice, however, the battery is well lagged in cellulose wadding inside the transmitter case and the risk of failure even in ascents to high altitude is small. Further protection is given by enclosing the battery and its lagging in a plastic bag, the main purpose of which is to prevent acid from spilling on to the transmitter when the instrument strikes the ground.

For the purposes of transit and storage the batteries are packed in dozens in sealed tinned-steel boxes with silica gel drying agent. To facilitate filling the cells with acid a polythene bottle with a suitable delivery tube (and a funnel for filling the bottle) and a polythene ampoule for removing excess acid are provided. Details of the preparation of a battery for operational use are given in the *Handbook of radio-sounding technique*[46].

8.2.6. The ground equipment

The apparatus for receiving and measuring the signals of the radio-sonde at the ground station consists essentially of a radio receiver, a calibrated variable audio-frequency oscillator and a cathode ray oscilloscope. They may be mounted in chassis with front panels designed to fit a standard 19 in. rack, as shown in Plate XXVI. The receiver output, in the form of an audio-frequency signal, is applied to one pair of plates of the oscilloscope and the output of the variable oscillator to the other pair. When the two frequencies are equal the trace on the screen takes the form of a stationary loop. The oscillator can be set rapidly by this means to within 0·1 c./s. of the frequency of the signal, the value of which is read from the oscillator dial. Calibration curves provided with the radio-sonde then enable the frequency readings to be converted into measurements of pressure, temperature and humidity.

Meteorological Office radio-sonde receiver, Mark 1.—Although some commercial and military types of radio receiver with suitable radio-frequency band coverage can be used with the Mark 2B radio-sonde there are advantages in using the Mark 1 receiver (Stores Ref., Met. 3999) which was specially designed for the purpose. This is a superheterodyne receiver covering the frequency range 26 to 30 Mc./s. The local oscillator is stabilized for voltage and temperature fluctuations and is fitted with a manual control. The local oscillator frequency is below that of the signal. An automatic volume control is provided, and also radio-frequency and audio-frequency gain controls and coarse and fine tuning controls. The sensitivity of the receiver is such that an input signal of 2 μV., 30 per cent. modulated, gives a signal-to-noise ratio of not less than 10dB. A high intermediate frequency of 1600 kc./s. is used in order to minimize the effect of unwanted image signals which might otherwise mask the relatively weak signal from a distant radio-sonde. The intermediate frequency has a band width of 20 kc./s. which is adequate to cover both the short-period shifts of radio-frequency that may occur during a sounding and the useful spectrum of side bands. Outputs are available at low impedance for operating a loudspeaker built into the receiver, and at a higher impedance for feeding into a separate oscilloscope display. The tuning dial is marked with graduations at every 0·1 Mc./s. and its calibration should be accurate to within

$\pm$ 0·05 Mc./s. The calibration can be checked at intervals of 0·5 Mc./s. by means of the harmonic output from a 500 kc./s. crystal oscillator which is incorporated in the receiver. A beat-frequency oscillator is also provided; this can be switched in when it is necessary to check that the radio-frequency carrier is being received during a failure of the audio signal.

In view of the importance of having an efficient aerial for the relatively weak signals, provision is made on the receiver for a switch enabling alternative forms of aerial, both tuned and aperiodic, to be used to suit different conditions. A ground-plane aerial tuned to the mid-frequency in the radio-sonde band generally gives the best results, but on the rare occasions when the transmitter is at a high angle of elevation a horizontal half-wave dipole or a conventional aperiodic inverted " L " aerial is more efficient.

Variable audio-frequency oscillator.—The oscillator developed for use with the Mark 2B radio-sonde is the Muirhead type D–207–D (Stores Ref., Met. 3200). It comprises a mains power unit, stabilized in its H.T. section, an oscillator section and an output amplifier. The oscillator section consists essentially of a two-valve amplifier with negative feed-back through fixed resistors and positive feed-back through a resistance–capacitance tuning circuit. It has a pre-set reaction control for adjusting the overall gain and, with automatic voltage control, produces the required constant amplitude of oscillation. The output from the oscillator circuit is fed to the output amplifier valve through a variable resistor, thus providing an output from a transformer wound to suit a 600 Ω impedance, which can be varied up to a maximum of at least 5 V. The tuning range covered is from 700 to 1005 c./s. and the scale is graduated in units of 1 c./s. round a large semicircle, on which the frequency setting can be rapidly made by means of a pointer with a combined direct and slow motion drive to which the tuning capacitors are coupled. Backlash effects do not exceed $\pm$ 0·1 c./s.

In the initial calibration the scale is adjusted to read correctly to within $\pm$ 0·2 c./s. throughout its whole range. Provided the a.c. mains supply is adequately stabilized the stability of the oscillator is such that the calibration does not drift by more than 0·2 c./s. per hour after a suitable warming-up period (of at least an hour), but it is necessary to check the calibration immediately before and after a sounding against a standard or a substandard frequency source. Provision is made for small adjustments of the calibration by means of a trimming capacitor or by re-setting the pointer. Instructions for these and for other adjustments are given in the makers' operating and maintenance manual supplied with the instrument.

Audio-frequency reference standards.—It is necessary that the frequency scales of the measuring oscillators used at the radio-sonde ground stations should be compared with the same reference standard as that used in the calibration of the radio-sondes. Two types of equipment are available for use as substandards in such comparisons. One of these is a valve-maintained tuning fork of the Muirhead type D–20–A (Stores Ref., Met. 3199). The tuning fork itself is made of elinvar, which has a very small temperature coefficient of elasticity, and provided it is operated within the usual range of room temperature no temperature corrections are necessary. The driving amplifier which maintains the fork in oscillation has a feed-back circuit and is designed to give a high degree of stability; a mains voltage change of 15 per cent. produces less than 1 part in 10^6 change in frequency. The fork frequency is 1000 c./s. and is initially adjusted by means of a pre-set control to within 20 parts in 10^6 of this figure. The maximum error from all causes other than temperature variation should be less than 30 parts in 10^6, while a temperature change of 1°C. produces a change in frequency of the order of 10 parts in 10^6.

The other type of substandard for audio-frequency is a quartz crystal oscillator of British Physical Laboratories type L.O.166 (Stores Ref., Met. 3875). In this instrument the output of a 100 kc./s. crystal-controlled oscillator is first converted into a square wave and then, by a further valve, into a triangular wave. This voltage is used to lock a series of multivibrators which divide the 100 kc./s. oscillation by 100 in four stages. The 1000 c./s. oscillation thus produced is passed through a tuned filter and then amplified. The frequency is accurate to within 5 parts in 10^6 at 20°C. and the temperature coefficient is less than 10 parts in 10^6 for 1°C. change.

Operating instructions for these instruments are given in makers' manuals supplied with the instruments. In using either of the substandards for calibration of the variable oscillator (before and after every sounding) the two audio-frequency outputs are applied to the appropriate plates of a cathode ray oscilloscope. The pattern on the screen depends on the ratio and relative phase of the two frequencies. If the ratio is expressed by simple integers a definite pattern, known as a Lissajous' figure, is formed and the figure is stationary. Examples are shown in Fig. 26. When the figure is stationary the ratio of the two frequencies is given by the number of loops of the figure that are tangential to the horizontal direction on the screen divided by the number tangential to the vertical direction. Thus with the variable oscillator set at 750 c./s. and the substandard oscillator at 1000 c./s. a ratio of 3/4 will be indicated by a figure with 3 loops along the horizontal and 4 along the vertical directions.

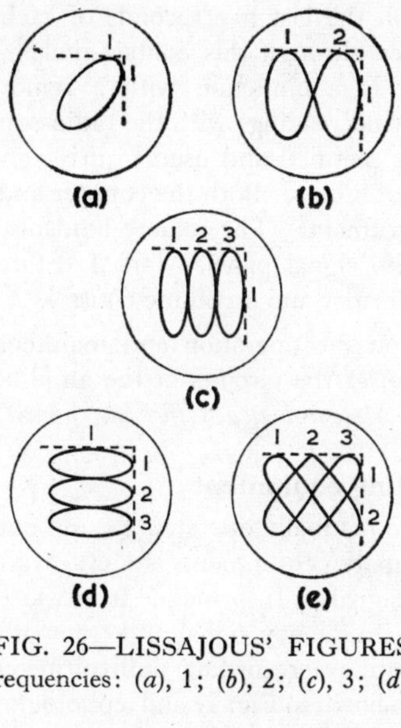

FIG. 26—LISSAJOUS' FIGURES
Ratio of frequencies: (*a*), 1; (*b*), 2; (*c*), 3; (*d*), $\frac{1}{3}$; (*e*), $\frac{3}{2}$

To ensure that the audio-frequency substandards used at the radio-sonde stations are in agreement with those used in the calibration of the radio-sondes, a weekly check should be made against an absolute standard of frequency. For this purpose the substandard frequency is compared, by means of the oscilloscope, with one of the standard frequencies that are broadcast regularly and can be received on a communication receiver. Details of the operational procedure are given in the *Handbook of radio-sounding technique*[46].

Meteorological Office radio-sonde oscilloscope.—Although the aural method of matching the pitch of the radio-sonde signal with that of a beat-frequency oscillator can be used, the visual method of comparison by means of a cathode ray tube is superior and is therefore adopted. Most commercial types of oscilloscope are suitable but are unnecessarily elaborate for the radio-sonde application and a simpler instrument has been specially designed. It is known as the Meteorological Office radio-sonde oscilloscope, Mark 2 (Stores Ref., Met. 10536), and in addition to the cathode ray tube display it provides timing data and signals for synchronizing radio-sonde and radar-wind operations.

The oscilloscope consists of a cathode ray tube of 6·3 cm. diameter, a pair of paraphase amplifiers to apply signals to the X and Y plates of the tube, and the associated power units. The amplifiers provide an overall peak-to-peak gain of about 150. While the gain of the amplifier fed from the variable frequency oscillator is fixed, a gain control is provided for the amplifier supplied with the radio-sonde signal to allow audible indication of the signal to be adjusted to a convenient level by means of the receiver gain control. A full-sized trace is produced on the oscilloscope screen by an input of about 2 V. r.m.s.

The equipment operates from 200/250 V., 50 c./s., a.c. supply and is provided with two internal d.c. power packs. One of these supplies 500 V. at 20 m. amp. to the amplifiers, and the other supplies 1500 V. at 5 m. amp. to the cathode ray tube. The usual controls for brilliance, focus and signal gain are provided.

For displaying elapsed time, in twentieths of a minute, a contact clock with a contact-maker operating on the last five seconds of each minute is provided. A signal of 1 kc./s. is applied through this contact-maker to a valve amplifier, the output of which is used in conjunction with a remote loudspeaker for aural synchronization of radar-wind readings with the radio-sonde observations. Part of the output of this valve is rectified and used to drive an electromagnetic counter displaying elapsed time in minutes. Both the counter and the clock can be reset to zero before using the instrument. The remote loudspeaker unit is known as the Meteorological Office audio signal panel, Mark 1 (Stores Ref., Met. 6017), and includes a matching transformer and a volume control.

Detailed information on the operation and maintenance of the oscilloscope, together with a description of the circuits of the amplifier, cathode ray tube and power supplies is given in *Meteorological Office instrument instruction* No. 252.

8.2.7. Automatic recording equipment

As an alternative to the manually operated ground equipment for the Mark 2B radio-sonde a fully automatic equipment is now available for measuring and graphically recording the signals. In principle it measures the time corresponding to 100 cycles of the audio-frequency signal and transforms the time into a representative voltage for operating a recording millivoltmeter. The general arrangement of the components is shown in Fig. 27 and a photograph of the whole apparatus is reproduced in Plate XXVII.

The received audio-frequency signal, after passing through a filter amplifier having a pass-band of 690 to 1020 c./s., is fed into a pulse shaper unit. This produces one positive and one negative going pulse for each cycle of input wave form.

These pulses are then fed into a 1 : 100 scaler unit which operates from the negative going pulse only. Thus one output pulse is produced for every 100 input pulses, i.e. 100 cycles of the input are contained between two such pulses. These pulses pass into a " seconds pulse unit " (virtually a line divider) which feeds a " start " pulse to a counter chronometer on receipt of the first pulse, and a " stop " pulse on receipt of the second.

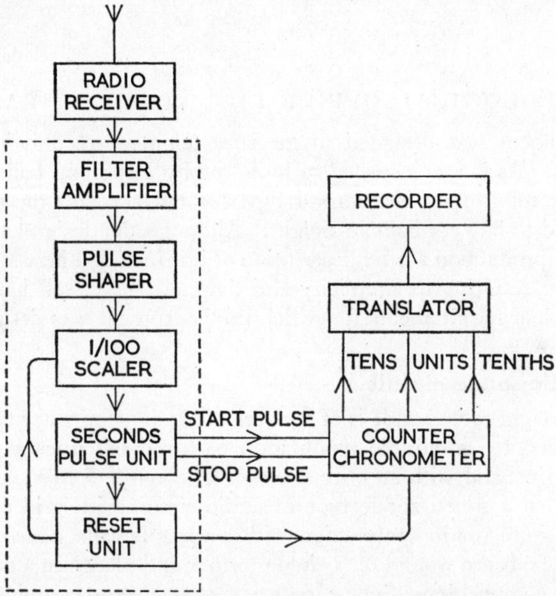

FIG. 27—BLOCK DIAGRAM OF AUTOMATIC RECORDING EQUIPMENT FOR METEOROLOGICAL OFFICE RADIO-SONDE, MARK 2B

The chronometer consists of a 100 kc./s. crystal oscillator, a gate and four decade counters in cascade, each associated with a meter scaled 0–9. When the gate is open individual cycles of the crystal oscillator are counted and when it is closed the total count is displayed. As each cycle corresponds to 10 μ sec. the dials record time directly in tens of microseconds. Thus on receipt of a start-and-stop pulse the chronometer will measure the time interval between them, i.e. the time for 100 cycles of the input frequency. This interval may vary between 0·10000 and 0·14286 sec., corresponding to the audio-frequency range 1000 to 700 c./s. of the signal, and as the measurement is made to within 10 μ sec. discrimination of frequency measurement of between 1/10 and 1/20 c./s. is obtained. Since there is little risk of ambiguity in omitting the first two figures of the count the last three figures only are fed from the chronometer to a translator which converts the decade information into a step-voltage wave form of 999 steps. This is used for operating a pen recorder which produces a graph of the last three figures of each count.

At the beginning of each count the equipment is automatically re-set to its initial position and four or five counts are recorded during the period of each audio-frequency signal. Interposed between each signal group there is usually one random noise count which coincides with the " no signal " period. The recorder chart is 25 cm. wide and the overall accuracy of the system is such that differences of 0·25 c./s. in the transmitted frequency can be resolved.

For most soundings completely automatic plotting can be obtained provided any drift in the sonde frequency is allowed for by re-tuning the radio receiver. If the signal-to-noise ratio is so poor as to make fully automatic operation impracticable, the recorder can be used for automatic plotting of frequencies measured by the normal manual operation of the measuring oscillator. With fully automatic operation no signals from the sonde are missed and a finer-structure record is produced giving greater detail over each signal period; this is illustrated in Plate XXVIII.

8.3. METEOROLOGICAL OFFICE LIGHT-WEIGHT RADIO-SONDE

This instrument was designed to measure temperature only to a minimum height of 8 km. As it was intended to be borne by a 100 gm. balloon which also carried a radar reflector of the shroud type for radar height measurements, the radio-sonde had to be very light in weight. Although the design has not been put into large-scale production a brief description of it is included here since some of its special features, resulting in simplicity and lightness, may well have applications beyond the limited requirements for which the instrument was designed.

8.3.1. The radio-sonde circuit

The light-weight radio-sonde is of the variable audio-frequency type, the signals being transmitted by amplitude modulation of a radio-frequency carrier on the 27·5–28·0 Mc./s. band with an output power of about 4·5 mW. It will be seen from Fig. 28 that a tuned anode type of audio-circuit with grid coupling coil is used. A fixed inductor in conjunction with a variable capacitance is used. The anode and grid coils are wound on a single former and placed in a pot-type core of a special ferrite composition. Such cores are very stable and combine high effective permeability with very low loss factors. The tuning capacitor placed across the anode coil is the temperature element which is in the form of a silvered, thin-walled, ceramic tube. The ceramic dielectric is barium strontium titanate which has a high temperature coefficient of permittivity and the five-fold change in capacitance of the unit over the temperature range + 40°C. to − 60°C. gives a frequency range of approximately 3000 to 7000 c./s. In order to eliminate " wet bulb " effects that might occur through wetting by rain or cloud droplets the capacitor is impregnated

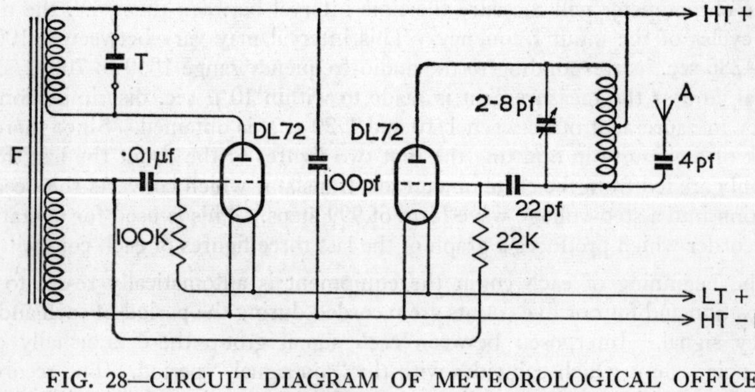

FIG. 28—CIRCUIT DIAGRAM OF METEOROLOGICAL OFFICE LIGHT-WEIGHT RADIO-SONDE

F is the ferroxcube assembly, T the capacitor temperature element and A the aerial.

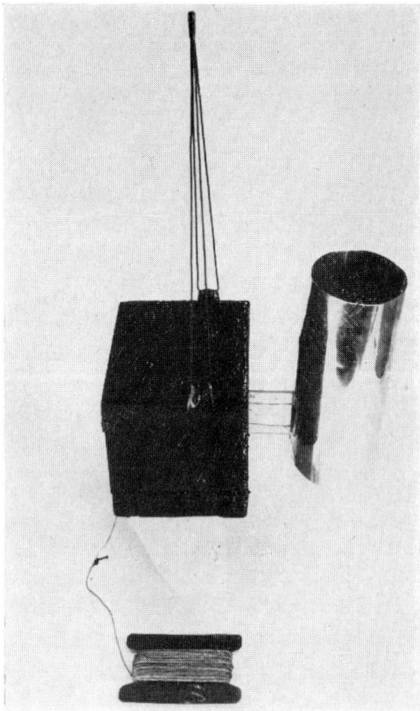

Complete radio-sonde

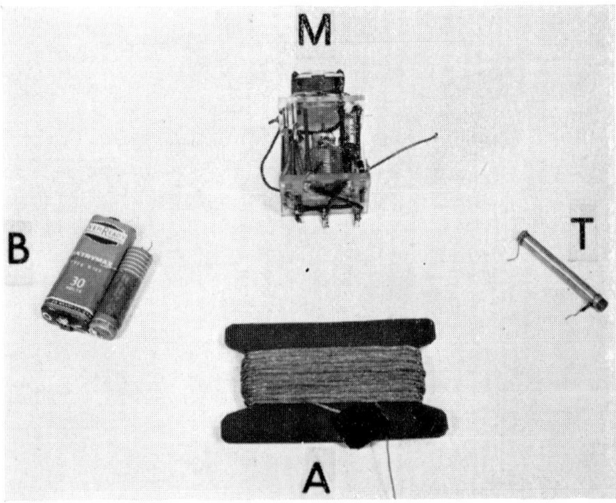

Components of radio-sonde

METEOROLOGICAL OFFICE LIGHT-WEIGHT RADIO-SONDE

A, aerial (wound); B, batteries; M, modulator/transmitter; T, temperature element.

Plate XXX

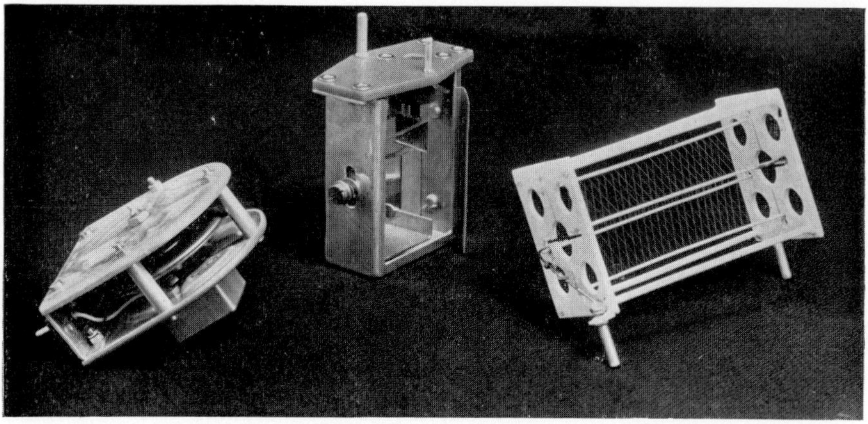

By courtesy of Mullard Ltd.

RADAR-SONDE METEOROLOGICAL UNITS

with a silicone resin. The grid coil of the audio-frequency oscillator is coupled to the base of the radio-frequency grid-leak resistor and thus grid-modulates the radio-frequency stage which is a conventional Hartley oscillator. A half-wave end-fed aerial completes the circuit. Power is supplied by two small deaf-aid batteries of 30 V. for H.T. and 1·5 V. for L.T.

8.3.2. Construction

The chief features of the construction are illustrated in Plate XXIX. All the components of the radio-sonde except the temperature unit are housed in a rectangular container made of expanded ebonite (onazote) 2·5 cm. thick. This material is a good heat insulator of low density and high rigidity. The container reduces the effects of temperature coefficients of components inside it to negligible proportions. It also maintains the batteries at a satisfactory operating temperature. The temperature element is mounted vertically 7·5 cm. away from the outer surface of the container and is surrounded by a double cylindrical radiation shield of aluminium. For calibration it is only necessary for the temperature element to be placed in the calibration chamber. The total weight of the radio-sonde complete with aerial and batteries is 270 gm.

8.3.3. Ground equipment

This is very similar to that used for the Mark 2B radio-sonde, except that the measuring oscillator covers the higher audio-frequency range of 2800 to 7500 c./s. A scale subdivided into 10 c./s. divisions is provided; readings to within 5 c./s. can be readily made. The receiver sensitivity required is about 3 : 1 signal-to-noise ratio for a 30 per cent. modulated carrier wave producing 2 μ V. at the receiver input with a 20 kc./s. I.F. band width. Both the audio-frequency 1000 c./s. substandard for checking the measuring oscillator and the oscilloscope for matching the transmitted and oscillator modulation frequencies are the same as those used for the Mark 2B radio-sonde.

8.4. THE RADAR-SONDE

An account has already been given in Section 5.2 of the secondary radar system developed for upper wind measurement by the Mullard Research Laboratory[55] to the design of Jones and his associates[54] in the Ministry of Supply. The complete system included the use of the 10·5 cm. transmission from the transponder for the additional purpose of telemetering the measurements of pressure, temperature and humidity. The performance of the radar-sonde developed for this purpose did not meet the operational requirements but, since the design included a number of novel features and is capable of further development, a short description is included here.

8.4.1. General principle

The method of telemetering is to transmit a pair of pulses in response to each pulse from the ground, the first pulse of the pair being the ranging pulse and the separation between the two being controlled by the meteorological units. The latter are switched into the circuit in turn in a time cycle of about 15 sec. Conversion of the wind-finding transponder into a complete radar-sonde therefore requires the addition of three meteorological units, a switch mechanism and a telemetering circuit for producing the pulses with the variable delays controlled by the meteorological measurements.

At the ground the 10·5 cm. transmission, with 404 pairs of pulses per second, is received by the radar-theodolite which, if it is to interpret the sonde signals in addition to the ranging signals, requires the addition of a telemetering unit. The principle of this is to measure the time delay between the pairs of pulses by counting the number of standard (1 Mc./s.) time intervals in the delay. This process is repeated for a total of 500 pairs of pulses during the period each meteorological unit is switched into the circuit. The first four figures of the total count are recorded digitally by a teleprinter.

The radar-sonde system is, in effect, a chronometric one in which the timing is done electronically; it is, therefore, capable of very high accuracy. The range of variation of the time delay between pairs of pulses is from 200 to 1200 μ sec., depending on the value of the meteorological variable. As the counting process is repeated 500 times the maximum possible count is 600,000 and any random error in the time delay is averaged out.

8.4.2. The meteorological units

Since the control of the pulse circuit is effected by voltage variations the most convenient type of meteorological element for the radar-sonde is the resistance type. Practical considerations, however, prevented the use of this type except in the case of the temperature element.

To produce the required resistance of about 6000 Ω a tungsten wire of 0·0125 mm. diameter is used for the temperature element. It is in the form of a coil (as used in electric lamps) and this is wound on a light frame, as shown in Plate XXX. Such an element is greatly superior to the bimetal type, both in response time, which is a small fraction of a second, and in radiation error. Further, the almost linear variation of its resistance with temperature simplifies calibration.

Both the pressure and humidity units, also shown in Plate XXX, operate with variable inductors with ferrite cores which are moved by the aneroid capsule and gold-beater's skin element respectively. The pressure unit is inside the thermally insulated container of the sonde and is therefore not subject to large temperature errors. The use of variable inductance instead of resistance for these units involves some additional complication of the pulse-delay circuit.

8.4.3. The pulse-delay circuit of the radar-sonde

The general arrangement of the main components of the airborne unit is shown in Fig. 29. To develop the variable delay between the ranging and meteorological pulses a phantastron circuit consisting of a pentode valve with a feed-back capacitor between the anode and the grid is used. When this is triggered by a pulse from the transponder modulating valve the anode voltage falls, at a constant rate, to a minimum value at which a rapid rise occurs and from which a pulse is obtained by differentiation. The range over which the anode voltage varies, and therefore the time delay between the triggering and derived pulses, is determined by the initial voltage and this is controlled by the meteorological units. The delayed pulse is fed into the modulator to trigger this, and the 10·5 cm. transmitter, a second time.

In the case of the temperature element the phantastron anode voltage is obtained directly by passing a suitable current through the wire resistance element. For the pressure and humidity units the inductance variations have to be converted into voltage variations. The inductor coil in each of these units forms part of the circuit of an oscillator, the variable frequency output of which is fed to a tuned filter. This produces an amplitude variation dependent on the inductance variation and the rectified output of the filter gives a voltage suitable for applying to the phantastron anode.

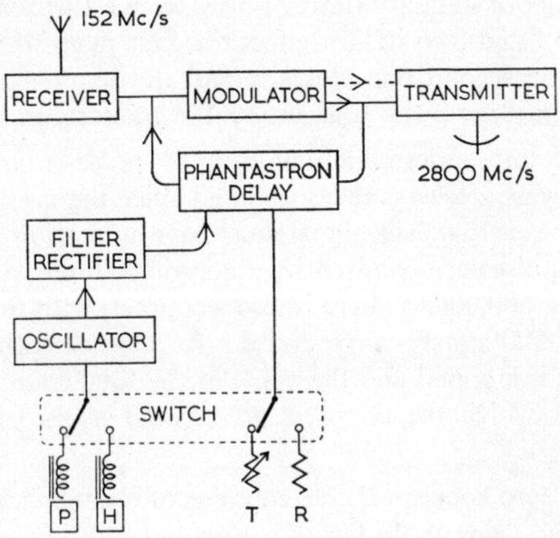

FIG. 29—BLOCK DIAGRAM OF MAIN COMPONENTS OF RADAR-SONDE AIRBORNE UNIT

P, pressure; H, humidity; T, temperature; R, reference resistor.

In addition to the three meteorological signals transmitted during each 15 sec. switching cycle two reference pulses, one derived from a fixed inductor and the other from a fixed resistor, are transmitted. These are used for correcting the pressure and temperature readings for circuit variations due, for example, to changes in battery voltage or temperature. Certain coding pulses, of two types, are also transmitted during every switching cycle. They are generated by switching fixed resistors into the phantastron circuit at delays shorter than the minimum (200 μ sec.) delay of the meteorological pulses. Their purpose is explained in Section 8.4.4.

8.4.4. The ground telemetering and recording equipment

A simplified diagram of the additional equipment necessary for the telemetering and recording of the sonde information by the radar-theodolite ground station is given in Fig. 30. The equipment is required to identify the received signals, to

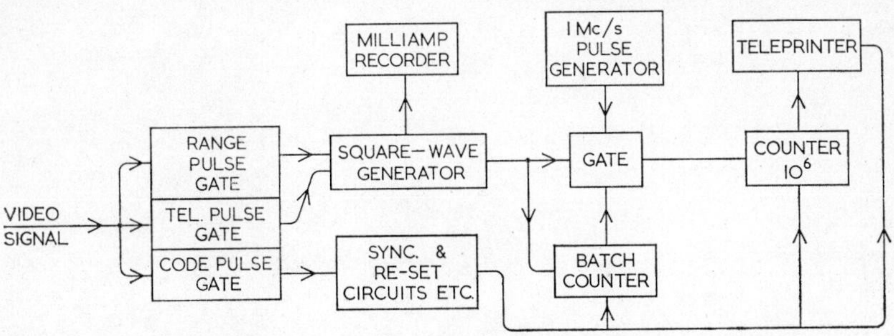

FIG. 30—BLOCK DIAGRAM OF RADAR-SONDE GROUND TELEMETERING UNIT

measure the time delay between ranging and meteorological pulses, and to record the results in correct sequence. A square-wave generator with a periodicity equal to the pulse repetition frequency of 404 per second is initiated into operation by the ranging pulse and cut off by the meteorological pulse, thus producing a square wave of length equal to the time delay between the pulses. This length is measured by counting the number of standard timing pulses from a 1 Mc./s. crystal-controlled oscillator that can be fitted into it. In effect the first pulse of each pair opens an electronic " gate ", the second one closes it and the number of standard timing pulses passing through it gives the time interval in microseconds.

In order to smooth out random fluctuations in the pulse-counting system a batch of 500 of the square-wave cycles is measured and since the cycles recur at a rate of 404 per second this means that each signal must have a duration of at least 1·25 sec. The 1 μ sec. timing pulses are counted by a decade counter with six decades, the maximum count being 600,000. There is also a counter, with three decades, which counts each batch of 500 square-wave cycles. As soon as each batch is counted, the main 10^6 counter is stopped and the value of the total count is transferred to a teleprinter unit, after which the counters are cleared to zero ready for the next signal.

The teleprinter record consists of five columns of four-digit figures, each representing the pulse time-delay in fifth's of a microsecond. From left to right the columns refer to pressure reference, temperature reference, pressure, temperature and humidity. The coding pulses generated in the airborne unit and transmitted between each of the five signals indicate that new information is about to be received and enable the counting circuits to be automatically re-set in readiness for it, while special coding pulses, which are transmitted once in every switching cycle, ensure that the teleprinter readings are maintained in the correct columns: if a signal is interrupted they restore the correct sequence of recording.

A graphical record of sufficient accuracy to show salient features of a sounding can be obtained by passing a constant current for the duration of the square waves through a recording milliammeter. The mean current so recorded is a measure of the time delay of the meteorological pulses.

CHAPTER 9

RADIO-SONDE CALIBRATION, BALLOONS AND ACCESSORIES

9.1. CALIBRATION

9.1.1. General

Ideally the manufacture of radio-sondes should be such as to produce instruments with a standard performance so as to make it unnecessary for them to be calibrated individually. Most designs fall short of this ideal and individual calibration is necessary. Moreover, even when radio-sondes can be reproduced with a standard performance, it is desirable that representative samples should be checked.

In general, the closer the calibration can be made to simulate flight conditions the better, but practical considerations may necessitate a departure from such conditions. The design of apparatus required depends largely on whether the complete radio-sonde has to be calibrated as a whole or whether the meteorological units can be tested separately; in the latter case considerably smaller apparatus can be used. It is an advantage if both pressure and temperature calibration can be done in the same chamber and if both quantities can be varied simultaneously at the desired rates. The range of control should be adequate to cover the readings likely to occur in actual soundings and so make extrapolation unnecessary. Cooling can be either by mechanical refrigeration, in which case propane is a suitable refrigerant for the range of temperatures needed, or by immersing the calibration chambers in liquid coolants such as trichlorethylene cooled by the addition of solid carbon dioxide. Adequate ventilation of the meteorological units should be provided in the chambers and there must be suitable means of measuring the pressure, temperature and humidity in the working space to accuracies within, say, $\pm$ 0·2 mb., $\pm$ 0·2°C. and $\pm$ 1 per cent. relative humidity.

Pressure units that are not fully compensated for temperature must be calibrated at more than one temperature in order to provide for some system of correction. Humidity units are generally calibrated in a separate apparatus. In one type a duct or chamber is supplied with a continuous flow of a mixture of air from two vessels, one kept saturated with water and the other dried by silica gel; the relative humidity is controlled manually by a valve regulating the flows into the duct. An alternative arrangement is a chamber in which air is circulated through one or other of a number of vessels containing saturated solutions of salts. Any one of the vessels can be put into the circulation system by means of a valve, thus giving relative humidities of known values.

9.1.2. Meteorological Office radio-sonde calibration plant

Meteorological Office radio-sondes are calibrated individually; for this purpose the meteorological units are detached from the sondes, and these units only are placed in the calibration baths. Each unit, however, remains connected electrically to its transmitter and the measurements, in terms of audio-frequency, are made on ground equipment similar to that used in the routine observations. The ranges covered are from surface to 50 mb. for pressure, from about + 45°C. to − 85°C. for temperature and from below 15 per cent. relative humidity to above 95 per cent. In the system adopted the calibrations for the three quantities are carried out separately.

Temperature calibration.—For this the thermometer units are placed in vessels which can be completely immersed in liquid baths maintained at different temperatures. The baths are cylindrical copper vessels 43 cm. in diameter and 89 cm. high containing about 45 l. of liquid. Water is used for the higher temperatures and trichlorethylene, with a freezing point of − 86·4°C., for the lower temperatures. Two water baths are thermostatically maintained at approximately + 45°C. and + 25°C. by electric immersion heaters, while five trichlorethylene baths are maintained to within ± 3°C. of + 5°C., − 20°C., − 45°C., − 65°C. and − 85°C. by the supply of crushed solid carbon dioxide. All the baths are insulated with either glass-wool or expanded rubber and only the surfaces of the liquid are exposed to the air temperature. To maintain an even temperature the liquids are agitated by bubbling air through them.

The vessels which are immersed in the baths are copper cylinders 33 cm. in diameter and 61 cm. high with a flange around the top. Each vessel houses the lower part of the apparatus, shown in Plate XXXI; to the framework B of the apparatus 18 thermometer units are plugged. When the vessel is clamped to the plate A, with a rubber gasket in between, it forms a sealed calibration chamber. The enclosed air is circulated by a fan C just below the plate and is directed on to the thermometer elements by means of two baffle plates D and a central celluloid chimney E. The latter houses two copper–constantan thermocouples for measuring the air temperature. Above the plate A is a central column on which the fan motor G is mounted and as the same design of head is used for pressure calibration a double pressure seal is fitted in the column H. The heat generated by friction at the seals, which are spring-loaded, is dissipated by filling the central column with oil which is kept in constant circulation by a small impeller on the spindle. A small tube J provides the return circuit for the oil and a reservoir K ensures that the seals are kept wholly immersed in oil. At the top of the apparatus two large Tufnol discs L carry the transmitters of the radio-sondes and these are connected to their temperature units by leads which pass through the four tubes M, through which the thermocouple wires also pass. Seventy leads pass through the tubes, all of which are pressure-sealed at the bushes N. A battery or stabilized mains supply of 85 V. high tension and 2·4 V. low tension is connected to the head, the L.T. supply passing through a uniselector switch O so that only one radio-sonde is connected at one time.

Plate XXXII shows the temperature calibration plant in use. After the head is loaded up the copper vessel is clamped in place and immersed in the + 45°C. bath. When the temperature inside the vessel has become steady the readings are taken, the thermocouple instrument being read to 0·1°C. and the audio-frequency measuring oscillator to 0·1 c./s. The apparatus is then transferred to the + 25°C. bath and its place taken by a second vessel, and so on, until all the baths are occupied by vessels, by which time there will be over 120 thermometer units in various stages of calibration. Small amounts of solid carbon dioxide are added to the baths as necessary to maintain the right temperature. A thermometer head and its vessel fully loaded weighs about 80 kg. and transfer from one bath to the next is done by means of an electric travelling hoist running on a gantry. Despite the heavy weight it is necessary to add a further 9 kg. of lead to sink the vessels in the trichlorethylene, which has a specific gravity of 1·48. The baths are provided with guides to keep the vessels upright and central.

Pressure calibration.—The vessels and heads used for pressure calibration are similar to those used for temperature calibration but, because of the smaller radiation shield, 30 pressure units can be accommodated in the apparatus. Two outlets on the heads allow suction lines from a rotary pump to be fitted and the pressure in the vessel is measured with a mercury manometer which is read to 0·2 mb. The pressure units are calibrated at 15°C. from surface pressure to 50 mb. with readings at nine points, but as the units are sensitive to temperature a calibration is also done at − 60°C. with readings at 300, 200 and 100 mb., thus giving the frequency changes due to the change in temperature at these pressures. This calibration at the low temperature is, in fact, done first.

Humidity calibration.—The hygrometer units are calibrated in a separate apparatus. This consists of a chamber in which air is rapidly circulated, by means of a small compressor, round a closed circuit, no outside air being drawn in. A frame holding 32 units is placed in the chamber and the humidity of the circulating air is controlled by diverting it, before it enters the chamber, through one of three canisters. One of these contains silica gel which dries the air to about 15 per cent. relative humidity. A second contains water heated to about 3°C. above room temperature by a small immersion heater, thus giving about 97 per cent. relative humidity. A third canister contains unheated water and by mixing air which has passed through this canister with air from the silica gel canister humidities ranging from 15 to 80 per cent. can be obtained, depending on the proportions going through each canister. Calibrations are generally made at about 97, 70, 40 and 15 per cent. relative humidity. Readings are taken on a psychrometer, the bulbs of which are within the chamber.

9.2. CONTROL CORRECTIONS

9.2.1. General

It is important that a check of the calibration of a radio-sonde at surface values of pressure, temperature and humidity should be made shortly before the instrument is used, since it is impossible to ensure that no changes occur during the interval, often of some months, between calibration and use, especially if the hazards of transport are involved. The departures from calibration values revealed by such a check may arise from a variety of causes such as:

(*a*) Drift in the properties of the meteorological elements and other components.

(*b*) The use of different standards of frequency at the calibrating and observing stations.

(*c*) The influence of connecting leads and other conductors used during calibration.

(*d*) The use of different battery voltages during calibration and control check.

The changes from any one cause may be small but the cumulative effect may be of importance. If the departures revealed by the control check are unduly large the radio-sonde may require adjustment or re-calibration, but for small departures corrections can generally be applied. For making the control observations some form of screen or chamber is necessary to ensure that the radio-sonde and the control instruments are measuring the properties of the same sample of air.

9.2.2. Meteorological Office control screens

The control check of a Meteorological Office radio-sonde consists in measuring the audio-frequencies indicated by the meteorological units for the surface conditions just before a sounding, and comparing these with the frequencies derived from the calibration curves. Unless the radio-sonde is defective the change in frequency is not likely to be more than 1 or 2 c./s. and it is applied as a constant correction to the calibrations. This procedure is not strictly correct since some of the factors may be expected to produce changes of different amounts at different frequencies. However, the approximation allows simple and rapid computation to be made and, in order to limit the errors that may arise from it, no radio-sonde should be used whose control corrections exceed 5 c./s. for temperature or pressure or the equivalent of 10 per cent. for humidity.

Two main types of control screen are in use at Meteorological Office stations. One type, the Mark 2, is for use outdoors but it is being superseded by an indoor control screen, the Mark 5.

Meteorological Office radio-sonde control screen, Mark 2.—This screen (Stores Ref., Met. 3334), which is illustrated in Plate XXXIII, consists of a steel housing, with double walls and roof, painted white. The radio-sonde is placed on a bracket about half-way up the back wall of the screen. An electric fan installed at the base of the screen draws air in between the walls and the roof, downwards past the radio-sonde, causing the windmill to rotate as in actual flight conditions. The aerial of the radio-sonde passes through an insulated slot in the back wall. Control readings are provided by an electrically driven Assmann psychrometer mounted so that its wet-bulb and dry-bulb thermometers extend through a side wall of the screen into the airstream just above the radio-sonde. An inspection window is fitted in the other side wall opposite to the psychrometer and internal electric lighting is provided by a fitting near the roof. The door of the screen is in two parts, the upper giving access to the radio-sonde and the lower to the fan. On the outside back wall are fitted a main switch, a switch and plug for the psychrometer and two other switches controlling the fan and the internal lighting. The screen is installed with the legs sunk about 50 cm. into the ground, and with the door facing north.

A disadvantage of an outdoor screen is that since the air temperature is often fluctuating rapidly, perhaps by as much as 1° or 2°C., it is difficult to ensure that individual readings or means of small numbers of readings are strictly comparable, for even if they are carefully synchronized the differences in lag between the psychrometer thermometers and the radio-sonde elements may have an appreciable effect in fluctuating conditions. The disadvantage is overcome, to a large extent, by the use of an indoor screen.

Meteorological Office radio-sonde control screen, Mark 5.—This indoor screen (Stores Ref., Met. 10511), illustrated in Plate XXXIV and Fig. 31, is in the form of a cube made of insulating panels fastened to an aluminium angle-section framework. A door on the right-hand side gives access to a sliding platform on which the radio-sonde is placed and below which can be inserted a flat cartridge of a sponge-type plastic. The latter, when saturated with water, enables high humidities to be produced within the screen. A deep short-bladed fan, rotating at 920 r.p.m. in a vertical plane, is situated on the left wall of the screen towards the rear upper part; it is driven by a 210/240 V., 50 c./s. motor fitted on the outside of the wall. The screen has an inner lining suitably curved to improve the air flow, which passes downwards past the meteorological elements, as in flight. Wet-bulb and dry-bulb thermometers are mounted behind a perspex window in the front panel of the screen and are illuminated by totally reflecting perspex rods which transmit light from a 60 W. source fitted outside the screen and controlled by a

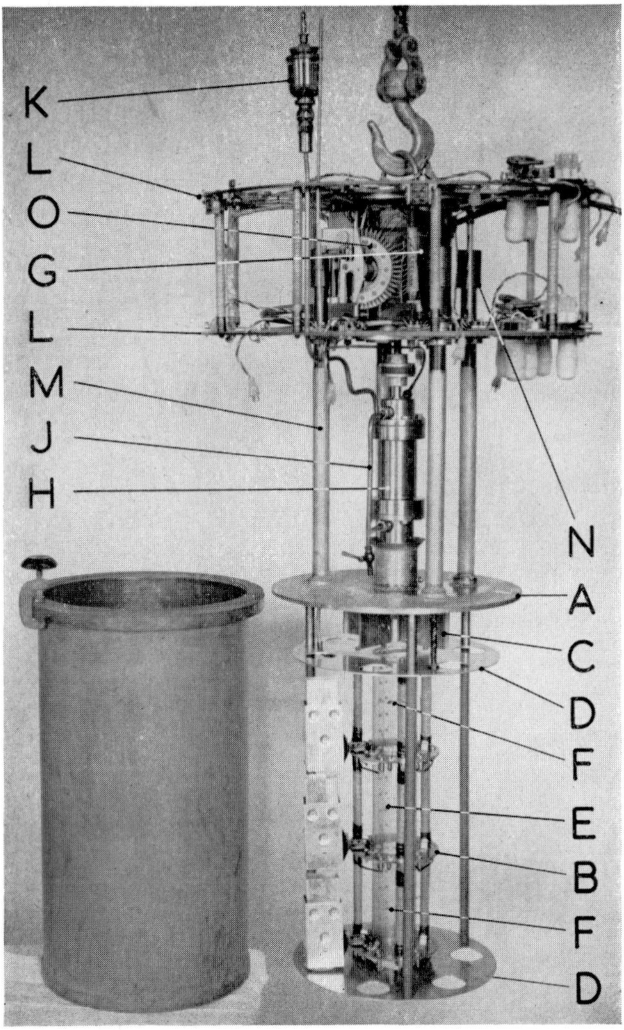

TEMPERATURE HEAD OF METEOROLOGICAL OFFICE RADIO-SONDE CALIBRATION PLANT

A, flat plate to form part of air-tight seal
B, framework holding 18 thermometer units (three seen here)
C, fan to circulate the air
D, baffle plates to direct the air flow on to the units
E, celluloid chimney
F, copper–constantan thermocouples
G, electric motor to drive the fan
H, column housing the double pressure seals
J, return circuit for the oil in the double pressure seals
K, oil reservoir
L, tufnol discs to carry the radio-sonde transmitters, three of which are seen here
M, connecting tubes carrying electric wires
N, pressure seals for tubes M
O, uniselector switch

Plate XXXII

METEOROLOGICAL OFFICE RADIO-SONDE CALIBRATION PLANT IN USE

METEOROLOGICAL OFFICE RADIO-SONDE CONTROL SCREEN, MARK 2

Plate XXXIV

METEOROLOGICAL OFFICE RADIO-SONDE CONTROL SCREEN, MARK 5

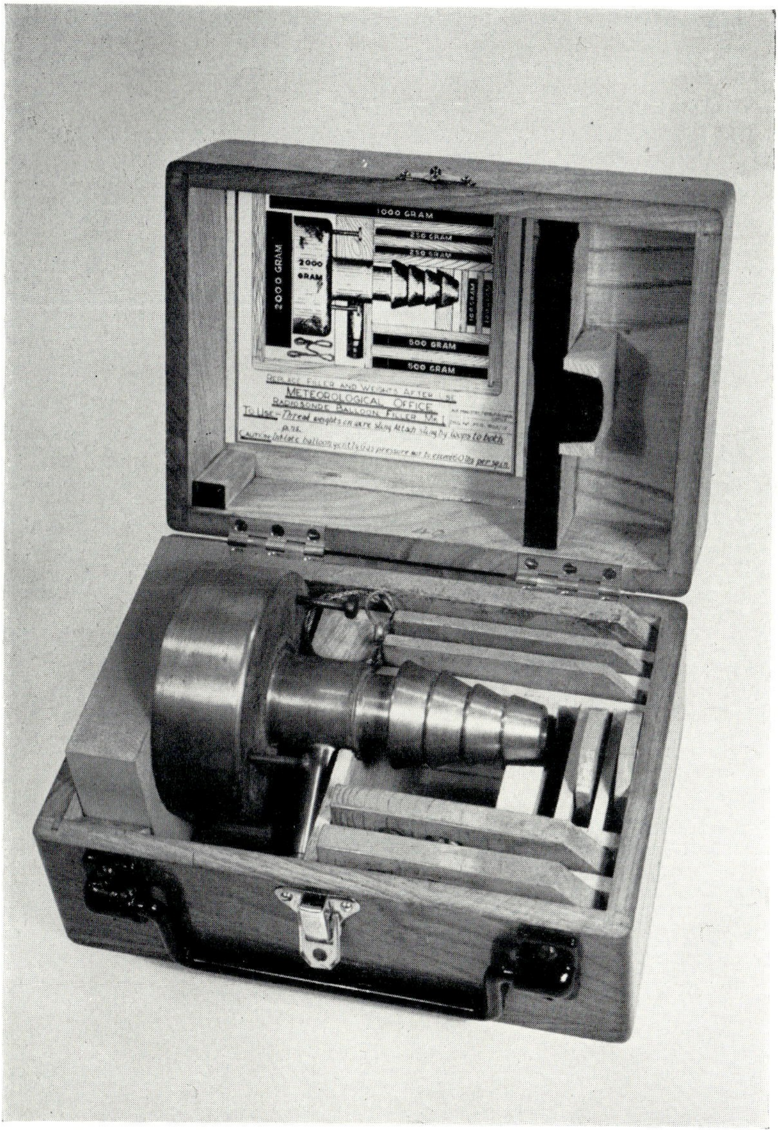

METEOROLOGICAL OFFICE RADIO-SONDE BALLOON FILLER, MARK 1

Plate XXXVI

METEOROLOGICAL OFFICE RADIO-SONDE STRING UNWINDER, MARK 2B

press-button switch. As only a fixed quantity of air is circulated in the screen, fluctuations in temperature are negligible and any temperature change is very slow (normally less than 0·05°C. per minute).

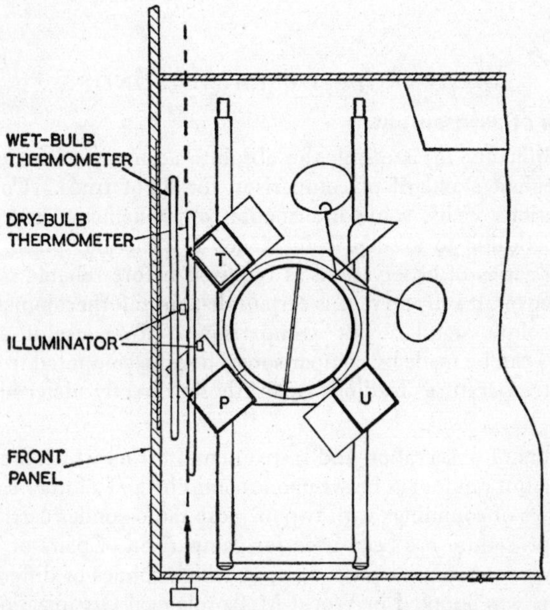

FIG. 31—PLAN VIEW OF METEOROLOGICAL OFFICE RADIO-SONDE CONTROL SCREEN, MARK 5, SHOWING SONDE IN POSITION

T, P and U are the temperature, pressure and humidity units respectively.

The screen is intended to be used on a table of normal height so that the thermometers are at a convenient level for a seated observer. In order that the temperature unit should be nearest the thermometer bulbs the meteorological units should be plugged into the radio-sonde in the order: humidity, pressure, temperature, in the clockwise direction from the windmill when viewed from above. The instrument is placed on the sliding platform, when this is pulled out, in such a position that the thermometer and pressure units are equidistant from, and close to, the front panel of the screen when the platform is pushed in. This positioning is most important as the clearances available are small; some adjustment of the windmill arms may, in fact, be necessary.

The humidifying cartridge should normally be stored in its case and only removed for use in its holder in the screen five or ten minutes before, and during the period of, the control determination. It should be wetted under a cold tap or by immersion in water not appreciably above room temperature. After use the cartridge must be returned to its case and the door of the screen left open, otherwise there is a risk of differential water absorption causing the insulating wall material to warp. No attempt should be made to raise the humidity in the screen above that produced by a single humidifying cartridge and the screen must not be used as a means of seasoning the gold-beater's skin hygrometer elements.

A distilled-water chamber, with a level-indicator on the front panel, is provided with a valve which should be opened for a sufficient time to wet the wick of the wet-bulb thermometer before each control observation. Care should be taken to prevent water dropping on to the wick during an observation. The procedure for making the control observations is given in the *Handbook of radio-sounding technique*[46].

9.3. ERRORS OF RADIO-SONDES

9.3.1. Methods of assessment

The main difficulty in assessing the absolute accuracy of radio-sondes is the lack of a convenient standard of comparison for flight trials. Comparison with aircraft observations yields some information on systematic errors but, as it is generally impracticable to arrange coincidence in time and place for such comparisons, a long series of observations is required before reliable mean values are obtained. Moreover, it is by no means certain that aircraft thermometers themselves provide an absolute standard for comparison. More useful assessments of systematic errors can be made by comparison of heights computed from radio-sonde pressure and temperature readings with those directly determined by radar measurements.

Random errors of observation and instrumental errors which are systematic for a given sounding but random as between one sounding and another can be estimated by making a series of soundings with two or more radio-sondes carried by the same balloon. This procedure has been used for comparison of pairs of Meteorological Office radio-sondes. A similar procedure, in which sondes of different types were sent up together, was adopted in World Meteorological Organization comparisons at Payerne, in Switzerland, in 1950 and 1956, the aim being to discover whether significant systematic differences occurred between the different types and also to estimate the scatter of individual soundings. Some of the results will be referred to in Section 9.3.3.

9.3.2. Sources of error

Although errors in radio-sonde measurements are more likely to arise in the airborne equipment (including batteries which may have voltage variations sufficient to cause errors in some types of sonde), there may be sources of error in the ground receiving and recording equipment. For example, errors in frequency measurement or in time measurement may occur, but they are not likely to be serious and are generally easy to estimate by comparison with standard laboratory test equipment. Automatic recording is more likely to introduce additional systematic errors, since it usually involves more elaborate mechanism than is necessary for manual operation, but it should reduce the random errors of observation.

Sources of pressure error.—Errors in pressure measurement may arise from drift in the calibration of the aneroid capsules, which is most likely to occur if the capsules have not been properly seasoned by being subjected to many pressure cycles through their working range. Such errors are reduced by the application of the corrections derived from the control readings. Hysteresis errors are not likely to be serious during ascent if calibration is carried out with pressure falling, but if observations during descent are used these errors may become large enough to require correction. Mechanical linkages in pressure units are liable to produce errors through backlash, but in a well designed unit these are not likely to exceed

0·5 mb. Unless the pressure unit is quite insensitive to, or completely compensated for, temperature variations, systematic errors may arise in the application of temperature corrections if the unit is at a different temperature from that measured by the radio-sonde. This will occur if the lag coefficients of the pressure and temperature units differ appreciably.

Sources of temperature error.—Unless the temperature unit is fitted with a thin wire resistor or a thermistor type of element it is likely to have systematic errors due to lag and radiation. The radiation error can, however, be reduced considerably by using elements with a low absorption factor and by efficient shielding. In its simplest form the error due to lag and radiation may be written:

$$S - T = \frac{Q}{q} - \frac{C}{q} \cdot \frac{dS}{dt},$$

where T is the true air temperature, S the observed value, Q the rate of absorption of radiation, C the heat capacity of the thermometer element and q the coefficient of heat transfer by forced convection. C/q is, in fact, the lag coefficient and the second term is the error due to lag, the first term being the radiation error. Besides being dependent on the form and size of the thermometer element, q increases with the mass rate of air flow and, since this is the product of the rate of ascent and the air density, both the radiation and the lag errors increase with height. There may also be an additional error at high levels if radiation absorbed by the shield is transferred by forced convection to the thermometer element. If the shapes of the latter and of the shield are not too complicated it is possible to calculate the coefficients of heat transfer and rates of absorption of radiation and so derive corrections for the radiation and lag errors. This has been done for Meteorological Office radio-sondes by Scrase[106, 107].

Other sources of error in the temperature unit are drift from calibration, which may not be entirely eliminated by the control corrections, and deposition of water or ice on the thermometer element, causing it to read low by the wet-bulb effect if the air is not saturated. A positive error is caused by the latent heat released if water deposited on the element freezes.

Sources of humidity error.—Humidity units in radio-sondes are much more liable to have large errors than either the pressure or temperature units. These errors arise mainly from the low rate of response of hygrometer elements at low temperatures and their insensitivity at high and low humidities. In some types of element, such as gold-beater's skin and ordinary hair, the errors become so large at − 20°C. as to make quantitative measurements useless. Gold-beater's skin and electrolytic resistor elements are also liable to produce large errors if they become wetted by rain or by long exposure to wet fog, and the electrolytic type may also be subject to errors caused by polarization.

9.3.3. Magnitude of errors

The World Meteorological Organization comparisons of 1950 and 1956 showed significant systematic differences between the readings of the various types of radio-sonde that were flown together. Results of the two series of comparisons are included in the final report of the W.M.O. Working Group[109] concerned with the comparisons. Although it is uncertain whether the differences found between different types of radio-sonde were characteristic of the types even at the time of the trials, it appears certain that type differences of several millibars in pressure and one or two degrees Celsius in temperature do exist. It is not known, however, how constant these type differences are over long periods of time. A further conclusion from the trials is that the scatter between individual radio-sondes of one type is roughly equal to that between one type and another.

9.4. RADIO-SONDE BALLOONS

9.4.1. General description

The requirements for sounding balloons for carrying radio-sondes and associated equipment differ from those of pilot balloons (see Section 3.1.3) mainly in the size of load to be carried, the height to be reached and the rate of ascent. Generally, the load, excluding the weight of the balloon itself, is between 0·5 and 2 kg., the heights required are from 20 to 30 km., or even 40 km., and the rate of ascent about 6 m./sec.

To meet these requirements in routine soundings extensible rubber balloons are nearly always used in preference to inextensible balloons. If an inextensible balloon is to carry the same load to the same height as a rubber balloon it must be as large as the latter at the point of bursting and must not weigh appreciably more. This means that the inextensible balloon must be made of very thin film, and it is only recently that plastic films sufficiently thin for the purpose, namely about 6 μ, have become available. A balloon made of such thin film, however, is very difficult to handle in any but very quiet conditions without some risk of it tearing or splitting, especially since it is only partially filled at the start. Moreover, its lack of rigidity gives it a poor aerodynamic shape which adversely affects the rate of ascent.

The rubber used in the manufacture of radio-sonde balloons is generally natural rubber or synthetic material such as neoprene, both of which are in the form of latex emulsion. This has to be compounded with small quantities of other materials and then vulcanized to produce a final product with the desired qualities of strength, extensibility and durability. The other ingredients include a sulphur-containing compound essential for the vulcanization process, an accelerator (usually a complex organic chemical compound) to speed up this process, an anti-oxidant such as a petroleum-type wax to prolong the useful life of the product by retarding changes due to atmospheric oxidation, and a specially developed organic chemical, termed a plasticizer, which increases the resistance to hardening at low temperature—a highly desirable feature in balloons to reach high altitudes.

There are two alternative processes by which the rubber is moulded into the form of a balloon. In one the latex emulsion is put inside a spherical mould which is immersed in hot water and revolved so that the latex is spread evenly and forms a gel over the inner wall. In the other process a mould is dipped into a coagulating agent and then into the latex which solidifies and forms a gel on the outer surface of the mould. In both methods the gel is removed from the mould, inflated while still wet and soft, then carefully dried and after deflation it is vulcanized. These methods have the advantage over the earlier method of construction from sheet rubber in that they produce balloons that are seamless.

9.4.2. Meteorological Office radio-sonde balloons

The balloons normally used at Meteorological Office stations are Beritex type, made of natural rubber, spherical in shape and uncoloured. They are specified by their nominal weight, and the actual weights of individual balloons may vary up to 10 per cent. of the nominal value. Neck lengths are generally between 10 and

11 cm. and the thickness of the unstretched rubber envelope is between 0·015 and 0·02 cm. except near the neck where there is a gradual increase of wall thickness to provide for supporting the required loads. As the rubber is compounded to have a very low modulus of elasticity and a high extensibility the internal pressure of an inflated balloon is very small, being only about 1 mb. above the outside air pressure, even up to the point of bursting, as found from tests at ground level. Other details of the balloons are summarized in Table X.

TABLE X—METEOROLOGICAL OFFICE RADIO-SONDE BALLOONS

Nominal weight	Stores Ref. Met.	Neck diameter	Unstretched diameter (approx.)	Minimum bursting diameter*
gm.		cm.	m.	m.
350	613	2·2	0·85	4·3
500	664	2·2	0·9	4·6
700	623	2·2	1·0	4·9
1250	1656	3·2	1·5	7·6
2000	870	3·2	2·0	9·0

* When inflated with air at ground level.

The balloons are supplied wrapped in cellophane and packed individually in sealed cardboard cartons, suitably lined. The sizes most commonly used are 500 gm. for soundings up to about 20 km. and 1250 gm. for soundings to between 25 and 30 km.

Care and handling.—The essential point in regard to the storage of radio-sonde balloons is that they should not be exposed to fresh air, light, or to extremes of temperature; where possible the temperature should be kept at about 68°F. The balloons should be left unopened in their cartons until they are about to be used.

If great care is taken at all stages of handling a balloon, pin-holes should rarely occur. Unless the floor of the filling hut is covered with linoleum or rubber and is kept quite clean the balloon should be placed on a piece of soft fabric for filling. This operation should be started very slowly and the fine-adjustment valve on the hydrogen cylinder should be adjusted so that the balloon will not be filled in less than fifteen minutes. The precautions to be taken in the use of hydrogen are given in Section 10.2.6 of Part 1. In order to prevent the generation of static electricity during the filling process both the hydrogen cylinder and the balloon filler must be earthed and, in addition, the balloon must be kept wet with soap solution.

The procedure to be adopted in launching a radio-sonde balloon is described in the *Handbook of radio-sounding technique*[46].

9.4.3. Rate of ascent

The rate of ascent of a balloon in still air is determined by the aerodynamic resistance D to the motion due to the lifting force Lg (L being the free lift in units of mass). When the rate of ascent reaches a steady value, which it does after a few seconds, D is equal to Lg. The resistance depends on the density ρ and dynamic viscosity μ of the air, the rate of ascent V and the diameter of the balloon d. It can be expressed in the general formula for the force which an airstream exerts on a body, thus:

$$D = \frac{\rho V^2}{2} \cdot \frac{\pi d^2}{4} \cdot C_D,$$

where C_D is the drag coefficient defined as the ratio of the resistance per unit area $(4 D/\pi d^2)$ to the dynamic pressure $(\rho V^2/2)$. C_D can be shown by the method of dimensions to be a function of Reynolds' number $V \rho d/\mu$ of the form:

$$C_D = C (V \rho d/\mu)^n,$$

where C and n are pure numbers, which must be determined by experiment.

The drag coefficient of solid bodies moving in a fluid is found to be relatively constant for Reynolds' numbers less than about 10^5 and more than about 3×10^5. In the critical range between these limits there is a large and rapid decrease in C_D (with increase in Reynolds' number) which is due to the transition from laminar boundary flow to turbulent flow. Data given by Goldstein[110] indicate that for solid spheres dropped from aircraft the drag coefficients are about 0·5 and 0·2 before and after the transition. For constant values of C_D the index n in the formula must be zero and the resistance is then proportional to the square of the velocity V.

The case of the ascending rubber balloon is rather more complicated than that of a solid sphere, partly because the diameter changes and partly because there is always some departure from the spherical shape which generally results in increased drag. The load carried by the balloon also adds to the resistance but, except in the case of a radar reflector of dimensions comparable with the balloon diameter, this extra resistance is probably negligible.

The total lift of the balloon is given by:

$$L + M = \frac{\pi d^3 \rho (1 - r)}{6},$$

where M is the combined mass of the balloon and the load and r is the ratio of the density of hydrogen to that of air at the same temperature and pressure. Combining this expression with that for D, the steady rate of ascent (for which $D = Lg$) may be expressed in the form:

$$V = \frac{b \mu^{(1-2/m)} L^{1/m}}{\rho^{(2/3-1/m)} (L + M)^{1/3}},$$

$$\text{where } b = \frac{8^{1/m} g^{1/m} (1 - r)^{1/3}}{6^{1/3} \pi^{(1/m - 1/3)} C^{1/m}} \quad \text{and } m = n + 2.$$

For constant values of the drag coefficient, when $n = 0$, the formula simplifies to:

$$V = \frac{b L^{1/2}}{\rho^{1/6} (L + M)^{1/3}},$$

and the rate of ascent then varies directly as the square root of the free lift and inversely as the cube root of the total lift and the one-sixth power of the air density. In pilot-balloon work the effect of the density is usually neglected and the formula for V becomes that given in Section 3.1.3, q being substituted for $b/\rho^{1/6}$. The value of q adopted for Meteorological Office pilot balloons up to the 30 gm. size, for which Reynolds' number is about 6×10^4, corresponds with a drag coefficient of 0·75, i.e. about 50 per cent. greater than for solid spheres.

In the case of radio-sonde balloons the effect of the density term is appreciable and causes a gradual increase in rate of ascent with height proportional to the following values of $(\rho_0/\rho)^{1/6}$, where ρ_0 is the air density at ground level:

Height (km.)	0	5	10	15	20	25	30
$(\rho_0/\rho)^{1/6}$	1·0	1·10	1·22	1·38	1·59	1·80	2·03

Unfortunately, however, the range of Reynolds' numbers applicable to these balloons generally includes the critical region in which there is the marked change in drag coefficient and therefore in the rate of ascent. This is shown in Fig. 32 in which some mean values of rates of ascent of three sizes of balloon and for various free lifts are plotted against height (measured by radar). The variations of drag coefficient with Reynolds' number calculated from some of these values are shown in Fig. 33. Up to at least 10 km. all the balloons have Reynolds' numbers greater than 5×10^5 and relatively low and constant drag coefficients of between 0·3 and 0·4. The increase of rate of ascent up to this height is almost entirely due to the density term in the formula for V and the rate of increase indicates that n in the general formula is approximately 2. Between 10 and 20 km. all the balloons, except the 0·5 kg. size with the higher free lifts, show a marked decrease in rate of ascent due to the rise in the drag coefficient (to about double the earlier value) which accompanies the fall in Reynolds' number into the critical region. Above 20 km. the drag coefficient of the 1·25 kg. balloons is again nearly constant at a

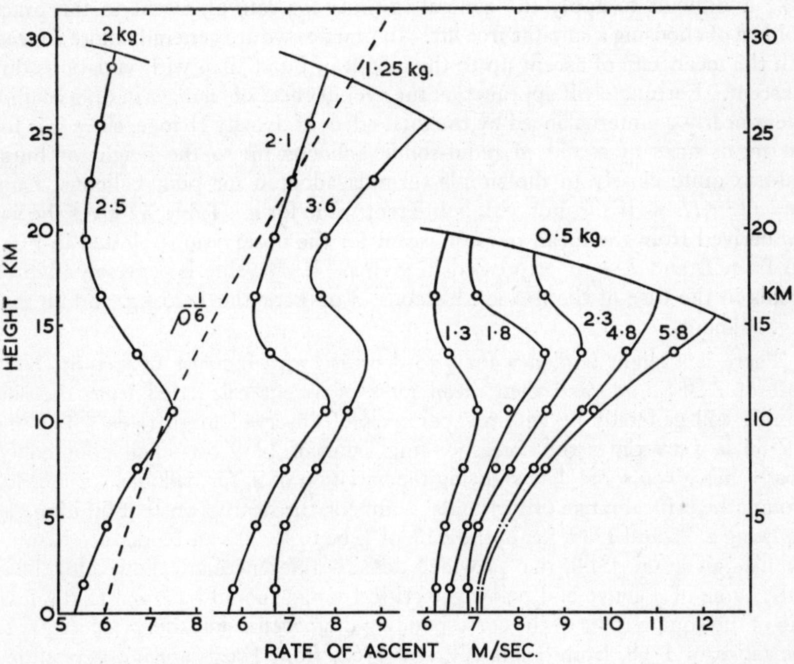

FIG. 32—VARIATION OF THE RATE OF ASCENT OF RADIO-SONDE BALLOONS WITH HEIGHT

Free lifts alongside curves. The 1·25 and 0·5 kg. balloons carried loads of 0·67 kg. The 2 kg. carried a load of 2 kg. The effect of the change of $\rho^{-\frac{1}{6}}$ on a rate of ascent initially 6 m./sec. is shown as a broken line.

value of about 0·65 and the further increase in rate of ascent is mainly due to the decrease in density. The drag coefficient of the 2 kg. balloon, on the other hand, continues to increase above 20 km., thus tending to counteract the density effect.

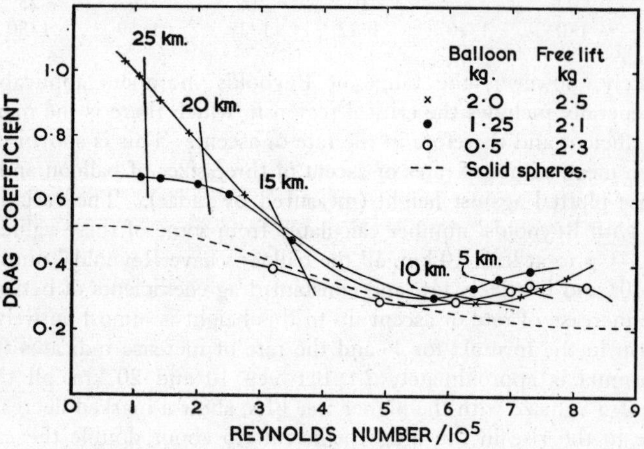

FIG. 33—VARIATION OF DRAG COEFFICIENT OF RADIO-SONDE BALLOONS WITH REYNOLDS' NUMBER

At first sight it would appear that these variations in drag coefficient would make it difficult to apply the general formula for rate of ascent to the practical problem of choosing a suitable free lift. In practice we are generally more concerned with the mean rate of ascent up to the height of burst than with variations during an ascent. Fortunately it appears that the overall effect of changes in drag coefficient is very nearly counterbalanced by the total effect of density change, since it is found that mean rates of ascent of radio-sonde balloons up to the height of bursting conform quite closely to the simple formula adopted for pilot balloons, namely, $V = qL^{1/2}/(L + W)^{1/3}$, but with a different value for q. Table XI gives the values of q derived from the mean rates of ascent for the observations plotted in Fig. 32, and from L and $L + W$ expressed in grams. Each value is representative of 30 ascents in the case of the 0·5 kg. balloons, 3 or 4 for the 1·25 kg. and 10 for the 2 kg. balloons.

There is a slight tendency for q to decrease with increase of free lift but if a value of 2·35 is adopted then mean rates of ascent calculated from the simple formula will generally be within 7 per cent. of observed mean rates. This figure of 2·35 is 1·69 times the corresponding value of 1·39 for small pilot balloons. Frost[111] has given some data showing the variation of q, for balloons without loads, through the critical range of Reynolds' numbers, the nearly constant limiting values of q being 2·25 and 1·39, i.e. in the ratio of 1·62 to 1. It can be deduced from the formulae given on p. 140 that provided density differences are allowed for the ratio of the value of q above and below the critical range should be equal to the inverse ratio of the square roots of the corresponding drag coefficients, i.e. $q_1/q_2 = (C_2/C_1)^{\frac{1}{2}}$. The ratios of 1·68, from Table XI, and 1·62, from Frost, are not very different from the value of 1·58 indicated by the ratio $(0{\cdot}5/0{\cdot}2)^{\frac{1}{2}}$ for the drag coefficients, quoted earlier, for solid spheres dropped from aircraft. Table XI also shows that the increase in mean rate of ascent becomes progressively less as the free lift is increased; no marked gain in rate of ascent is achieved by increasing the free lift beyond two or three times the combined weight of the balloon and load.

TABLE XI—OBSERVED MEAN RATES OF ASCENT OF RADIO-SONDE AND RADAR-WIND BALLOONS

Balloon weight	Free lift	Total lift	Height of burst	Mean rate of ascent	*q*
kg.	kg.	kg.	km.	m./sec.	
0·5	1·3	2·5	20·0	6·8	2·55
0·5	1·8	3·0	19·1	7·4	2·50
0·5	2·3	3·5	18·9	8·0	2·49
0·5	2·8	4·0	18·4	8·4	2·50
0·5	3·8	5·0	17·6	8·6	2·39
0·5	4·8	6·0	16·7	8·7	2·26
0·5	5·8	7·0	15·9	8·8	2·20
					Mean 2·41
1·25	2·1	4·1	28·6	6·8	2·36
	3·6	5·6	25·7	7·8	2·32
	5·1	7·1	21·1	8·6	2·30
					Mean 2·33
2·0	2·5	6·5	30	6·0	2·30

In Meteorological Office radio-sonde ascents with 0·5 kg. balloons carrying a load of 1·57 kg., a rate of ascent of about 6 m./sec. is aimed at and should be obtained by using a free lift of about 1·2 kg. For radar-wind ascents with the same size of balloons carrying a load of 0·665 kg., a free lift in the range 1·3 to 4 kg. may be used, depending on the strength of the prevailing upper wind which may necessitate a higher rate of ascent than 6 m./sec. if the balloon is to be kept within radar range. In combined radio-sonde and radar-wind ascents at 6 m./sec. with 0·5 kg. balloons a free lift of about 1·7 kg. is generally used. These free lifts are increased if precipitation is occurring at the time of launching, by 0·25 kg. in moderate drizzle and by 0·75 kg. in continuous moderate rain. A balloon that departs markedly from the spherical shape or is flabby after inflation may also require some modification of free lift.

9.4.4. Height performance

It is clear from Table XI that the higher rates of ascent can only be obtained at the expense of maximum height. For the 0·5 kg. balloons an increase of 1 kg. in free lift reduces the maximum height by about 0·8 km. Usually a reasonably high rate of ascent is required partly in order to provide sufficient ventilation for the radio-sonde and partly in order to keep the time required for a sounding within reasonable bounds, especially in a radar-wind ascent in which a strong wind may carry the balloon beyond the range of the radar. A compromise between the height requirement and the rate of ascent is therefore necessary.

The maximum height depends mainly on the total lift and on the size and extensibility of the balloon. The latter can be expressed as the ratio e of the diameter at bursting to that of the unstretched balloon. A bursting test at ground level, however, does not necessarily indicate the performance attainable in flight from balloons of the same batch since such factors as low temperature and the action of ozone and ultra-violet light may reduce the extensibility of the rubber. Moreover, in flight the load which the balloon has to support imposes extra strain on the rubber, causing some deformation from the spherical shape. The value of e obtainable from a balloon that gives a satisfactory performance in flight should be not less than four, and a good balloon should give a five-fold extension.

The factors determining maximum height are related by the buoyancy equation:

$$L + M = \frac{\pi d^3 \rho (1 - r)}{6},$$

where d is now the diameter at burst and r is again the ratio of the densities of hydrogen and air, assumed to be at the same pressure and temperature. If d_0 is the unstretched diameter we can then write the air density at the maximum height as:

$$\rho = \frac{6(L + M)}{\pi d_0^3 e^3 (1 - r)}.$$

Further, if we distinguish between the mass of the balloon M_b and the load M_s (together making M), then since $M_b = \pi d_0^2 t \sigma$, where t and σ are the thickness (generally between 0·015 and 0·02 cm.) and density (0·93) of the unstretched rubber of the balloon, we have:

$$\rho = \frac{6 \pi^{1/2} t^{3/2} \sigma^{3/2} (L + M_b + M_s)}{e^3 M_b^{3/2} (1 - r)}.$$

The extensibilities of balloons of different sizes and with different free lifts can therefore be compared by means of this formula.

The minimum bursting diameters given in Table X for balloons inflated with air at ground level correspond with a linear extension of about five. In actual soundings the performance corresponds with a slightly smaller extension. Thus for the data given in Table XI the values of e average 4·7 for the 0·5 kg. balloons, 4·3 for the 1·25 kg. and 4·4 for the 2 kg. The annual average performance of 0·5 kg. balloons at all Meteorological Office stations is summarized in Table XII.

TABLE XII—AVERAGE PERFORMANCE OF 0·5 KG. BALLOONS

Type of sounding	Free lift (approx.)	Total lift (approx.)	Height of burst	Linear extension ratio
	kg.	kg.	km.	
Radio-sonde	1·5	3·6	19·0	4·7
Radar-wind	2·4	3·6	17·5	4·3
Combined R.S.W. ..	2·0	4·5	17·0	4·5

The free lifts used were generally larger than are now recommended. There is a marked seasonal variation in the performance, the maximum heights being about 10 per cent. higher in summer than in winter. Differences also occur between night and day soundings, the heights being about 10 per cent. higher by day than by night.

In the theoretical discussion of balloon performance it has been assumed that the pressure and temperature inside the balloon are the same as for the ambient air. This is not strictly the case. In the day-time the effect of solar radiation is to raise the temperature inside by about 30°C. and this probably accounts for the better performance by day since, at the very low air temperatures of the stratosphere, rubber tends to lose its extensibility. The pressure inside the balloon always exceeds that of the ambient air by a small amount p' produced by the tension T of the rubber, p' being equal to $4\,T/d$. This sets a limit to the external pressure, and therefore to the height, which can be reached. This limit can be derived in the following way. The free lift at any height may be written:

$$L = M_a - M_h - M\,,$$

where M is again the mass of the balloon and load, and M_h and M_a the masses of the hydrogen and of the air displaced. Assuming the same temperature inside and outside the balloon L may be written:

$$L = M_h \left\{ \frac{p_a}{(p_a + p')r} - 1 \right\} - M\,,$$

where p_a is the air pressure and r the ratio of the molecular weights of hydrogen and air (or of their densities at the same pressure and temperature). Thus L decreases as p_a decreases, but the change is not appreciable until p_a falls to the same order of magnitude as p'. In the limit, L becomes zero and the balloon ceases to rise. At this point:

$$M = M_h \left\{ \frac{p_a}{(p_a + p')r} - 1 \right\}.$$

At the ground, where L_0 is the measured free lift, p' is negligible compared with the external pressure and so:

$$L_0 + M = M_h \left(\frac{1}{r} - 1 \right).$$

Eliminating M_h from these two equations gives:

$$p_a = \left\{ \frac{M + r\,L_0}{(1 - r)\,L_0} \right\} p' = \left\{ \frac{M + 0{\cdot}07\,L_0}{0{\cdot}93\,L_0} \right\} p' \approx \frac{M}{L_0} \cdot p'\,.$$

For some radio-sonde balloons p' has been found to be about 1 mb. at bursting point at the ground and, with M/L_0 about 2, the limit of height is that corresponding to a pressure level of 2 mb., i.e. 40 km. At low temperatures it is possible that the tension of the rubber increases and the height limit may therefore be less than 40 km. Heights of 45 km. have, however, been reached in a few soundings with 2·5 kg. balloons.

9.5. ADDITIONAL EQUIPMENT FOR RADIO-SONDE ASCENTS

9.5.1. Radio-sonde balloon filler

The principle adopted for the Meteorological Office radio-sonde balloon filler Mark 1 (Stores Ref., Met. 2819) is the same as that of the pilot-balloon fillers described in Section 3.1.3. The device, illustrated in Plate XXXV, consists of a brass casting incorporating two nozzles, a smaller one for connexion to the hydrogen supply and a larger one for attachment to the neck of the balloon. Two steel pins let into the casting form points of attachment for a detachable wire sling by means of which auxiliary weights may be hung from the filler to increase the weight. The casting weighs 2000 gm. and the detachable weights, which are made of lead, are 2000, 1000, 250 and 100 gm. These should be checked from time to time in case pieces of metal have been chipped or rubbed off in the course of use.

9.5.2. Parachutes

For the purpose of reducing the rate of descent of radio-sonde and radar-wind equipment to a safe value, paper parachutes (Stores Ref., Met. 10170) are normally used at Meteorological Office stations. These parachutes are made of four 60° segments of a circle of 90 cm. radius. The sides are bound together with tapes which extend about 1 m. below the base of the parachute. A strong cord extends through the centre from about 1 m. above to 1 m. below the parachute and is used to take the strain in the ascent, the upper end being attached to the balloon and the lower end to the remainder of the airborne assembly. The four corner tapes, which do not carry any of the main load, are used to maintain the parachute partly open so as to ensure its efficient opening during descent. For a radio-sonde ascent the tapes are tied to a metal or wooden hoop about 50 cm. in diameter and then continue to a common junction with the centre cord. The hoop is omitted when a radar reflector is carried, the tapes then being attached to the corners of the reflector The lengths of the tapes are adjusted so that they are slack, the load being taken by the central cord.

The material used for the parachute segments is kraft paper, crêped and waxed for waterproofing. It is cut so that the direction of the crêping is from the apex to the centre of the base of each segment. The total weight of the parachute is 165 gm. Plastic sheet is also being introduced for parachutes; it has the advantages of needing very little fabrication and of being waterproof.

9.5.3. String unwinder

In order to avoid the possibility of the radio-sonde being affected by the wake of the ascending balloon the instrument should be suspended about 40 m. below the balloon. This adds to the difficulty in launching the apparatus in strong winds (and in confined spaces as in ocean weather ships) and in such conditions a device known as a string unwinder Mark 1B (Stores Ref., Met. 10195) is used. It enables a short suspension to be used during the launching and then allows sufficient string to be paid out to lower the radio-sonde to its correct position 40 m. below the balloon.

The unwinder, which weighs 100 gm. and is shown in Plate XXXVI, has a spindle carrying a bobbin at its lower end and the suspension string is wound partly on the spindle and partly on the bobbin. The spindle is prevented from rotating by attaching it with a clip provided either to a parachute hoop (Stores Ref., Met. 10495) which is fitted with a vertical vane acting as an air-brake or to a radar reflector. An aluminium frame, also fitted with a vane to act as an air-brake, is carried by the spindle and tension on the string threaded through the frame causes it to rotate about the spindle, slowly at first until the string on the spindle is unwound, then more rapidly as the string comes off the bobbin, and finally slowing up towards the end of the unwinding. The driving force for this motion is provided by the weight of the radio-sonde. During launching there is normally sufficient wind acting on the vane of the unwinder to overcome this force and prevent the frame from rotating until the whole assembly is airborne.

9.5.4. Launching cone and tower

The difficulties of launching radio-sonde balloons in high winds lie mainly in handling the balloon on the ground and in preventing the radio-sonde assembly from being dragged along the ground when the balloon is released. These difficulties are reduced by using a launching cone and tower. In the Meteorological Office design of launching cone, Mark 2 (Stores Ref., Met. 10402), which is illustrated in Fig. 34, a protecting container is provided in which the balloon is carried. It may also be hoisted to the top of a tower sufficiently high to allow the radio-sonde assembly to hang vertically and then be released by a special mechanism.

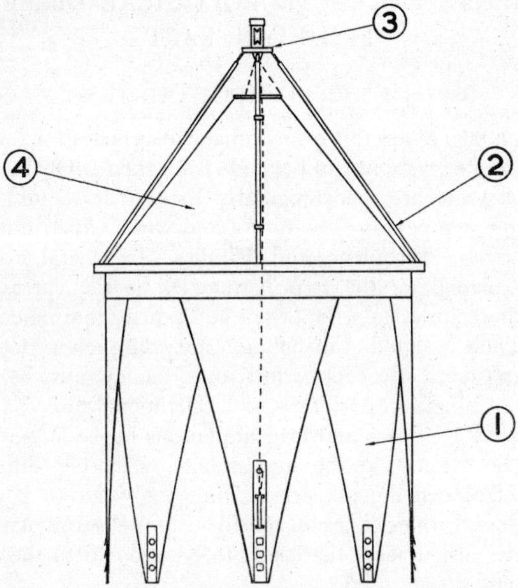

FIG. 34—METEOROLOGICAL OFFICE RADIO-SONDE LAUNCHING CONE
1, canvas sector; 2, strut; 3, trigger release; 4, canvas cone.

The container has a frame in the form of a pyramid made by attaching four tubular metal struts to a wooden hoop about 2 m. in diameter, their other ends being fixed to a small disc carrying a catch mechanism. Inside this structure is a canvas cone whose open base is fastened to the hoop while the slightly truncated top, also open, is attached by cords to the catch in the apex of the pyramid, in such a way that the canvas is held extended. Eight sectors of canvas have their bases permanently attached to the hoop and their tips can be drawn together by a cord passing through rings and held by a secondary catch. When this is released the sectors open like petals of a flower.

The balloon is placed in the cone and is held there by drawing the sectors together. In handling outdoors the apex of the cone is kept into wind, so that the canvas protects the balloon from wind pressure. The assembly is hoisted to the top of the launching tower and when the apex of the pyramid strikes the top the catches are released, thus freeing the top of the cone and the tips of the sectors. Pressure on the external surface of the cone, which eventually turns inside out, forces the balloon out and as the radio-sonde assembly is already vertically extended the apparatus immediately rises from the ground at the same rate as the balloon.

The tower is of angle-steel lattice construction, 12 m. high and 3 m. square at the base, tapering to 0·25 m. square at the top. A movable arm supporting a pulley for raising the cone is mounted at the top. Further details on the operational use of this equipment are given in the *Handbook of radio-sounding technique*[46].

CHAPTER 10

INSTRUMENTS FOR METEOROLOGICAL OBSERVATIONS FROM AIRCRAFT

10.1. GENERAL REQUIREMENTS

The main advantage of aircraft over unmanned vehicles for upper air measurements is that visual observations can be made from them. Of the aircraft meteorological observations that are made regularly those that require instruments are pressure and height, temperature, humidity and wind. Instruments may also be used for measuring icing, turbulence and visibility. In general, instruments for use on aircraft are designed on the same principles as the corresponding surface instruments but there are certain important additional requirements to which they must conform. Their design and installation must be such as not to endanger the safety of the aircraft or to affect its performance. They must be capable of withstanding vibration, changes in position and accelerational forces, and of being easy to read and manipulate. Space and load limitations in the aircraft usually impose size and weight limitations on the instruments. Another difference from the corresponding surface instruments arises from the need to cover much wider ranges of measurement. Special design considerations of aerodynamics and strength in the case of some instruments that need to be exposed to the outside air arise when high-speed aircraft are used.

In the early days of aircraft observations automatic recording by means of meteorographs mounted on the wings or struts was commonly employed, but with the increase of flying speeds such instruments became unsuitable and gave way to eye observations with direct or remote indicating devices. More recently, however, the use, for high-altitude measurements, of small fast aircraft with no accommodation for special air observers has again emphasized the need for automatic recording though now by electronic or other up-to-date methods.

Two main types of operation are used for aircraft meteorological observations. One takes the form of a roughly vertical sounding similar to a radio-sonde ascent, while in the other the chief aim is a horizontal reconnaissance. In the latter case, however, the observations on the horizontal track may be supplemented by vertical soundings at a few points, the soundings being made either by the aircraft or by a radio-sonde dropped from the aircraft. Drop-sondes designed for this purpose are described in Section 7.3.2.

10.2. AIRCRAFT INSTRUMENTS FOR PRESSURE AND HEIGHT MEASUREMENTS

For measuring atmospheric pressure from aircraft, instruments of the aneroid principle are used. An altimeter of the aneroid type may also be used if the law of calibration of the instrument is known, but since the other meteorological observations are usually related to levels of pressure, rather than height, it is more convenient to employ an aneroid barometer indicating pressure directly in millibars. The special requirements which distinguish the aircraft aneroid instrument from its surface counterpart are the wide range of pressure to be covered, the capability to withstand vibration, and the provision of an air-tight case. The latter is required since the pressure of the air inside an aircraft is not necessarily the same as that outside and the instrument must therefore be connected by a pipe to a point where as true a measurement as possible of the atmospheric pressure can be obtained.

10.2.1. Meteorological Office aircraft aneroid barometer

Description.—The instrument in use in Meteorological Office aircraft is the Mark 2B model (Stores Ref., Met. 663). It is designed to cover the range 1050 to 150 mb. on the type of dial shown in Plate XXXVII. The dial, which is about 7·5 cm. diameter, is divided into units from 0 to 100, the units being millibars for the long pointer and tens of millibars for the short pointer; the pointers, which are concentric, move counter-clockwise with decreasing pressure. The case of the instrument is made of moulded bakelite and is air-tight, except for the connecting nipple at the back, the toughened glass front being sealed with a rubber ring.

A double capsule is used as the sensitive element and the mechanism linking it to the indicator is illustrated in Fig. 35. The deflexion of the capsule C is transmitted through the jointed link LA to a shaft S to which a toothed sector H is fixed. This sector is geared through an intermediate pinion I to another pinion J which moves the long pointer P. Gears driving the short pointer are housed in a recess behind the dial-mounting and a screw for adjusting the pointers is provided; this screw, which is covered when it is not in use, meshes with gear teeth round the circumference of the dial-mounting. A hairspring attached to the staff of the intermediate pinion removes backlash from the mechanism.

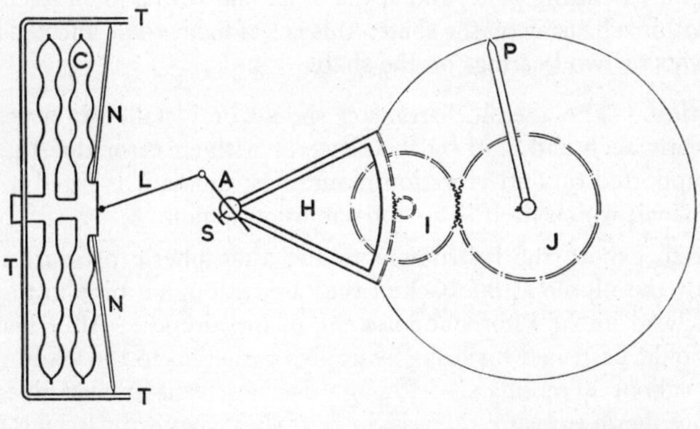

FIG. 35—MECHANISM OF AIRCRAFT ANEROID BAROMETER

C, capsule; L and A, connecting links; S, shaft (perpendicular to paper); H, sector; I and J, pinions; P, long pointer; T, bimetal strips; N, steel pins.

With the form of indication used in this barometer, rotation of the pointers must be linear with respect to pressure change over the whole range. The aneroid capsule has a nearly linear deflexion characteristic and the slight departure from linearity at low pressures is compensated by arranging for the magnifying lever and gear mechanism to have a suitable departure from linearity in the opposite direction. As the characteristics of individual capsules vary from one to another some adjustment is necessary, when the instrument is first assembled, to match the individual calibration. This is done by adjusting the effective length of the link A and also the distance between the capsule and the shaft S. The effect is to alter the working position of the mechanism curve, thus enabling a suitable part of the curve to be selected to match the individual capsule.

As the capsule has a temperature coefficient of elasticity a temperature-compensating device is fitted. This consists of a bimetal strip T bent at right-angles at each end so that it can exert pressure on the deflecting boss of the capsule through two steel pins N. The device is designed to give correct compensation at all pressures in the following manner. If the temperature increases at surface pressure the decrease in elasticity will tend to make the capsule collapse, but this is compensated by expansion of the bimetal strip outwards, thus decreasing the pressure of the steel pins on the capsule. At low atmospheric pressure the effect is the same but the movement of the capsule is smaller and the angle of the pins is less, with a correspondingly smaller effect. By suitable adjustments of the bimetal strip and of the positions of the pins, accurate compensation can be obtained over the wide range of temperature and pressure likely to be encountered.

In order to maintain the mechanism in balance in all positions of the instrument the shaft S is linked to a spring-hinged weighted arm (not shown in Fig. 35). By adjusting the tension of the spring and the length of the arm compensation is obtained for any change of reading due to the weight of the capsule in different positions. The balance weight also keeps tension on the links and so helps to eliminate any oscillation of the pointer.

The capsule assembly is designed to have a resonant frequency high compared with the frequencies of vibrations most likely to occur in an aircraft. Vibration errors are also reduced by mounting the shaft S with a pivot at one end fitting into a bearing in the mounting plate, and at the other end with a pivot fixed to the plate and fitting within a bearing on the shaft; this is less liable to be affected by vibration than two pivots or two bearings on the shaft.

Installation.—The aneroid barometer should be installed in a position where it can be clearly seen and read by the observer without error due to parallax. It should be supported on anti-vibration mountings, unless it is fitted to the aircraft instrument panel, which itself is an anti-vibration mount.

In order to expose the instrument to the atmospheric pressure it should be connected, by the nipple at the back of the case, through a pipe to the static vent, or the static head in the Pitot static system, of the aircraft. After installation the apparatus should be tested for leaks by applying suction to the static vent or head equivalent to about 30 millibars. Although the static vent or head gives very nearly the true atmospheric pressure there is, in fact, always an error, which may amount to a few millibars, known as the position error. This depends on the airspeed and the altitude of the aircraft and it is normally expressed, as a pressure error dp, in terms of a correction dv_i to the indicated airspeed v_i thus:

$$dp = -\rho_0 v_i dv_i ,$$

where ρ_0 is the air density at which the airspeed indicator is calibrated. The procedure for computing and applying the position-error correction is given in the *Meteorological air observer's handbook*[112], which also includes an appendix on the measurement of aircraft speed.

Calibration.—Aircraft aneroid barometers must be calibrated at three-monthly intervals and also whenever there is reason to doubt their serviceability and accuracy. For the calibration a vacuum test chamber and a mercury test barometer are required. Readings of the aneroid and the test barometer are made at room temperature with pressure both falling and rising, at the rate of roughly 30 mb. per minute. If the first reading at surface pressure differs from that of the test barometer the error may be eliminated by adjusting the pointer-setting screw before the rest of the calibration is continued. The allowable errors are as follows:

(*a*) The mean error at a given pressure for pressure rising and falling must not exceed 2 mb. at any point on the scale.

(*b*) The difference between the error for pressure rising and falling must not exceed 5 mb. at any point.

(*c*) The difference in readings at surface pressure for horizontal and vertical positions of the aneroid barometer must not exceed 2 mb. (The instrument should normally be calibrated in the attitude in which it is used.)

A leak test should be carried out in the test chamber by connecting the nipple of the instrument case to the external atmosphere and raising the pressure in the chamber to one atmosphere above the external pressure, when there should be no change in the pointer reading. Any instrument failing to satisfy this test and the calibration requirements should be replaced.

Errors.—Aircraft aneroid barometers are subject to the same sources of error as surface aneroids, viz., inadequate temperature compensation, hysteresis, friction, backlash, and secular change. These are discussed in Section 2.3.4 of Part I. In addition, aircraft aneroids may have errors due to lack of balance in the moving parts of the mechanism and to leaks in the instrument case. Position errors, though not arising in the instrument itself, also affect the readings, as explained above.

With the temperature compensation provided in the Mark 2B aneroid any residual error due to temperature effect should not exceed 1 mb. for a change of 10°C. Errors due to hysteresis should be within the tolerance of 5 mb. allowed in calibration. Their effect is to cause the indicated pressure to be greater than the true pressure when the aircraft is climbing and less than the true pressure when it is descending, by an amount depending on the rate of climb or descent. In order to reduce the effect and obtain a steady reading level flight is necessary for at least one minute, or considerably longer if there has been a rapid change of level.

Some typical values of errors of Mark 2B aneroids are summarized in Table XIII, which is based on test calibrations of six instruments.

TABLE XIII—MEAN ERRORS OF SIX AIRCRAFT ANEROID BAROMETERS, MARK 2B

	Pressure in millibars					
	1000	700	500	300	200	150
	millibars					
R.M.S. error of readings ..	1·2	1·3	1·0	1·1	2·1	2·8
Mean difference with falling and rising pressure	2·6	3·5	2·9	1·9	1·0	—

It will be seen that the mean error at any point was less than 2 mb. except at the two highest levels, and the maximum difference between readings with pressure falling and pressure rising was, on the average, less than 4 mb. The mean change of reading for a change of 10°C. was less than 0·5 mb.

10.2.2. The use of an aircraft altimeter for pressure measurement

A sensitive aneroid altimeter, such as one that meets the British Standard Specification 2G.115 of design and performance requirements, may be used for the measurement of pressure. This type of instrument is designed to indicate the height of the aircraft in accordance with the International Civil Aviation Organization (I.C.A.O.) standard atmosphere[113] which superseded the International Commission for Air Navigation (I.C.A.N.) relationship between atmospheric pressure and height to which such instruments were formerly calibrated. Since the maximum difference between corresponding values in the I.C.A.N. and I.C.A.O. conventions over the range 1000 to 150 mb. is only about 0·05 mb. the revised relationship has little effect on the design and calibration of the altimeters.

The design, which is exemplified in the Royal Air Force altimeter, Mark 14, shown in Plate XXXVII, provides for high magnification of the aneroid-capsule deflexion by a mechanism similar to that of the Meteorological Office Mark 2B aircraft aneroid barometer, and the dial, which has multiple pointers, can be read easily to 25 feet. A detailed description of the design and construction has been given by Hoather[114]. The sensitive altimeter is provided with an adjustable sub-scale in millibars over the normal range of surface pressure. This sub-scale is so related to the main height scale and gearing that it indicates at any setting the pressure in millibars at which the pointers will read zero height. Thus if the sub-scale is set to the surface pressure at an airfield the height indicated on the main scale will be that above the airfield provided no change of atmospheric pressure takes place during the flight. The accuracy of the sensitive altimeter is about the same as that of the Mark 2B aneroid.

The method of determining the pressure from the altimeter reading is described in the *Meteorological air observer's handbook*[112]. The I.C.A.O. standard atmosphere, on which the calibration of the altimeter is based, assumes a temperature of 288·16°K. and a pressure of 1013·250 mb. at mean sea level, and a tropospheric lapse rate of 1·9812°C. per thousand feet. The relationship between pressure (in millibars) at height (in thousands of feet) at other levels up to the assumed level of the tropopause (226·32 mb.) is:

$$P = 1013{\cdot}25\left(1 - \frac{1{\cdot}9812}{288{\cdot}16}H\right)^{5{\cdot}2561}.$$

If the datum pressure at which the altimeter is set to read zero height is not 1013·25, but P_s, then the indicated height is:

$$P = P_s\left\{1 - \frac{1{\cdot}9812}{288{\cdot}16}H\left(\frac{1013{\cdot}26}{P_s}\right)^{0{\cdot}19027}\right\}^{5{\cdot}2561}.$$

As this formula involves both P_s and H the computation is best done by means of a family of curves or a specially designed slide-rule. If, however, an altimeter is used frequently for pressure measurement it is more convenient always to set the sub-scale to 1013·25 and use a single curve or table representing the I.C.A.O. relationship. Such a table is given in Appendix VII of the *Meteorological air observer's handbook*[112] and a more detailed one is included in *I.C.A.O. Document* 7488[113].

10.2.3. Instruments for height measurement

Barometric altimeter.—The instrument described in the preceding section is primarily intended for height measurement. The instrumental accuracy of the sensitive altimeter made to the British Standard 2G.115 is such that in tests over the full range the errors must not exceed 70 ft. at zero height, 450 ft. at 25,000 ft. and 750 ft. at 50,000 ft. As, however, the instruments are designed to indicate height in a standard atmosphere, departures from the assumed temperature distribution may result in errors up to $\pm$ 10 per cent. A temperature correction should therefore be applied if greater accuracy is required. This can be done by deducing the pressure from the indicated height by the method described in the preceding section and, from a knowledge of the temperature distribution, evaluating the true height by using, for example, a tephigram. For navigational purposes, however, it is more convenient to use a Royal Air Force height and airspeed computer as described by Bull[115].

The aircraft aneroid altimeter is normally connected to the static pressure vent or head and is, therefore, subject to position error for which a correction must be applied. In terms of altimeter height this is given (in feet) by:

$$dH_i = + \frac{v_i \, dv_i}{11 \cdot 0 \, \sigma},$$

where dv_i is the correction to the indicated airspeed v_i, and σ is the relative air density at I.C.A.O. height H_i. The latter is tabulated in the *Meteorological air observer's handbook*[112], Appendix VII.

Radar altimeters.—Meteorological observing aircraft may be equipped with a radar altimeter for measuring the geometric height above land or sea surface. The surface acts as a plane reflector normal to the incident beam and with this favourable condition, together with the requirement to measure only range and not direction, a relatively simple radar system is adequate. Two types of equipment are in use, one employing frequency modulation and the other pulse technique. Both types are often referred to as radio rather than radar altimeters. As they are not primarily for meteorological use only brief mention will be made of them.

In the frequency-modulation instrument a continuous carrier wave modulated with a frequency between, say, 400 and 450 Mc./s. is transmitted from the aircraft. The signal reflected by the ground or sea surface returns with a frequency differing from that of the carrier at the instant of arrival of the return signal and the difference is a direct measure of the height above the surface. The difference is measured by a receiver, the output of which is fed into conventional shaping and counting circuits which operate a d.c. meter calibrated directly in height. This type of instrument is used for heights up to about 4000 ft. and can usually be read to within 10 ft., but the overall accuracy may be considerably less than this.

The other type of altimeter, using pulse technique, adopts the more usual radar principle of measuring the time delay of an echo pulse from the reflecting surface. The delay is measured on a cathode ray tube with a suitably extended time-base. Heights up to the order of 40,000 ft. are covered with this type of equipment and readings can be made to the nearest 50 ft. as a rule.

10.3. AIRCRAFT THERMOMETERS

10.3.1. General

In the early days of meteorological observing from aircraft the thermometers used for air temperature measurement were of the liquid-in-glass type. The aircraft were generally slow-moving biplanes and one of the struts provided a convenient mounting for the thermometer from which it could be seen by the pilot or observer in the cockpit. To enable the instrument to be read from that position it was either made very large or was provided with a magnifying lens which could be moved along the scale by a wire or cord controlled from the cockpit. With the disappearance of struts this type of installation became impracticable but a more suitably mounted liquid-in-glass thermometer continues to be used in relatively low-speed aircraft. For high-speed aircraft remote indicating electrical thermometers in which only a small sensitive element is exposed outside the aircraft are now used.

Most aircraft have air thermometers installed as standard equipment. Generally these are of the bimetallic or electrical type but are not of sufficient accuracy for meteorological requirements. In addition to the general requirements for aircraft instruments the thermometers for meteorological observations are now normally required to cover a range of about + 40°C. to − 80°C., to have an accuracy not inferior to ± 0·5°C., to be sufficiently rapid in response to give satisfactory observations in moderate rates of climb and descent, and to possess aerodynamic properties which permit satisfactory allowance to be made for the effects of aircraft speed.

This last requirement is perhaps the most important since, with the high aircraft speeds that are now commonplace, the effects of heating by compression of air at the thermometer element or by friction with air flowing past it are very large. These effects are discussed in detail in the next section.

A general account of thermometer lag is given in Section 3.1.6 of Part I and the effect of lag on aircraft observations of temperature is discussed in the *Meteorological air observer's handbook*[112]. It should suffice here to recall that the lag coefficient is inversely proportional to the mass rate of air flow past the thermometer element, i.e. to ρv, raised to a power between 0·5 and 0·8 depending on the form and dimensions of the element. It therefore increases with increase of height (decrease of ρ) and with decrease of airspeed.

Aircraft thermometers may be used in the form of a psychrometer but the special requirements of the wet bulb will be dealt with under hygrometers in Section 10.4.

10.3.2. The effect of aircraft speed on temperature measurement

When there is rapid relative movement between a thermometer and the ambient air the thermometer indicates a temperature which is higher than the true air temperature by an amount depending on the square of the relative speed. The heating arises from the motion in two ways, namely by compression of the air at the thermometer surface and by friction with the air flowing past it.

The dynamic pressure at the surface of a body moving through the air is, in general, different from the atmospheric pressure at the same level. Near the nose of the body there is an excess of pressure, while towards the rear this excess decreases and may become a deficit. These pressure changes take place adiabatically and are accompanied by corresponding temperature changes. At the surface of the body the relative motion between it and the air in contact with it is zero and this results in a velocity gradient in the boundary layer of the air which gives rise to frictional heating. The amount of frictional heat generated increases with the relative speed of the body in the air. Since, in accordance with Bernouilli's theorem for a stream-line in fluid flow, the pressure is low where the velocity is high, and vice versa, at points on the body where there is a pressure deficit the velocity at the outer edge of the boundary layer tends to be relatively high and thus frictional heat will counteract the adiabatic cooling due to the pressure deficit.

At the stagnation point on the nose of the body or inside a Pitot head, where the relative velocity is zero, there will be adiabatic heating by compression but no frictional heating. Since air effectively behaves as a perfect gas all the kinetic energy of motion is converted into heat and the temperature rise in the air at the point is given by:

$$\Delta T = \frac{V^2}{2 J C_p},$$

where V is the true airspeed in cm./sec., J the mechanical equivalent of heat ($4{\cdot}2 \times 10^7$ ergs/cal.), and C_p the specific heat of air at constant pressure (0·24 cal./gm.°C.).

When, however, air flows parallel to the surface of the body, as in the case of a flat plate parallel to the flow, there is no change of pressure and the generation of heat is due to friction only. This, however, is offset to some extent by forced convection of the heat from the boundary layer to the free airstream. The net rise of temperature, therefore, is determined by the balance between the viscous forces and the thermal conductivity of the air. This balance is governed by the Prandtl number $\sigma = \mu C_p/K$, where μ is the dynamic viscosity and K the thermal conductivity of the air. It can be shown that the temperature rise differs from that given above for adiabatic compression by a factor $\sigma^{\frac{1}{2}}$ when the air flow over the surface is laminar and $\sigma^{\frac{1}{3}}$ when it is turbulent. In a medium, therefore, for which σ is unity the full equivalent of the kinetic energy appears as heat. For air, σ has a value between 0·71 and 0·72 over the range of temperature and pressure in atmospheric conditions up to at least 30 km. So the temperature rise of a body such as a flat plate parallel to the flow is 0·85 times the full adiabatic rise, $V^2/2JC_p$, when the air flow is laminar and 0·89 times that rise when the flow is turbulent.

The above considerations of the effect of airspeed suggest two fundamental forms of aircraft thermometer. In one the air is arrested completely and the thermometer experiences the full effect of compressional heating. The other form of thermometer is one which disturbs the air flow as little as possible and the temperature rise depends on the balance between frictional heating and forced convection. The first form is generally known as a total-head thermometer, while the second form is exemplified by the flat-plate type. Alternatives to the latter may take the form of a cone or a thin cylinder with a rounded or pointed nose, with the axes in the direction of air flow. In earlier types of aircraft thermometer, however, no attempt was made to approach any of these forms.

In most practical cases precise computation of the temperature rise due to airspeed is complicated by lack of knowledge of the actual pattern of the air flow and of the dissipation of the excess heat in the turbulent wake of the body. Some temperature changes may, in fact, take place in the air before it reaches the thermometer surface, depending on where the thermometer is mounted on the aircraft. In general, therefore, it is necessary to determine the temperature corrections for aircraft speed by flight tests of the thermometer mounted in the position in which it is to be used. The aircraft is flown round a small circuit at a constant pressure level where the air is free from clouds and convection effects, and temperature readings are taken over a range of various known airspeeds. Since the temperature rise is proportional to the square of the true airspeed a graph of the readings plotted against the square of the true speed should give a straight line, from the slope of which the speed correction coefficient α_D can be derived for use in obtaining the dry-bulb temperature correction at any speed from the formula:

$$\Delta T = \alpha_D (V_k/100)^2 ,$$

where V_k is in knots and ΔT in degrees Celsius. The order of magnitude of the correction is indicated by an example in which α_D is unity thus giving a correction of 25°C. for an airspeed of 500 kt.

The theoretical value of α_D for the full adiabatic effect may be calculated from:

$$\alpha_D = \frac{100^2 V_k^2}{2 J C_p V^2} ,$$

and is 1·34. For the full frictional heating of a flat plate in laminar flow it is 0·85 of this value, viz., 1·14. For practical use the values derived from experimental determinations are given in the ensuing sections dealing with the aircraft thermometers in use.

When an aircraft thermometer is passing through cloud or rain some of the heat produced dynamically is used up in evaporating the cloud droplets or raindrops approaching the dry bulb. The speed correction is therefore reduced, the amount of the reduction depending on the liquid water content of the air as well as the airspeed. There is considerable difficulty in obtaining the dry-bulb readings in these conditions since the liquid water content is not, in general, known precisely. The problem is discussed in more detail in the *Meteorological air observer's handbook*[112] but it is best overcome, so far as is practicable, by making the temperature observations in clear air rather than in rain or cloud.

10.3.3. Meteorological Office aircraft psychrometer, Mark 6

Description.—This instrument, a photograph of which is reproduced in Plate XXXVIII, is designed for use in meteorological reconnaissance flights in multi-engined aircraft, on which it can be mounted outside the observer's window a few inches clear of the skin of the aircraft. Two versions of the instrument are available, the Mark 6A (Stores Ref., Met. 667) covering the range + 80° to − 60°F. and the Mark 6B (Stores Ref., Met. 675) covering the range + 110° to − 30°F. (the Fahrenheit scale being in use when the instrument was designed).

The dry- and wet-bulb thermometers fitted in the instrument are of the alcohol-in-glass type, with bulbs of flat oval cross-section. The thermometers are mounted in open-fronted brass tubes which are screwed into sockets soldered to the top plate of the bulb housing. To reduce the risk of breaking by vibration the thermometers are spring-mounted and are held in position against the springs by means of caps which screw into the top of each tube. The bottom half of the bulb housing acts as a water container for the wet-bulb supply, while the top half is designed as a radiation shield for the bulbs, the sides of this half being in the form of a honeycomb grille for ventilation. A sliding door is provided in the front of the housing to give access to the bulbs. The whole of the metal frame is nickel-plated and polished on the outside.

High accuracy is assured by the use of thermometers certified by the National Physical Laboratory to have errors at a point not exceeding 0·2°F. above 32°F., 0·4°F. between 32° and − 38°F., and 1·0°F. below − 38°F. Also the maximum change of error in an interval of 10°F. must not exceed 0·2°F. below, or 0·1°F. above, 32°F.

The lag coefficient of the dry-bulb thermometer, expressed as the time required for a temperature difference to fall to $1/e$ of its initial value, is 4·5 sec. at 200 kt. near ground level and it varies with air density and airspeed in accordance with the relationship $\lambda = 1/(\rho v)^{0 \cdot 7}$.

This type of aircraft psychrometer has the merit of simplicity, reliability and accuracy and although it is not suitable for use on modern high-speed aircraft it may continue to serve a useful purpose in certain types of aircraft that are still in use for meteorological observations.

Installation.—The instrument should be mounted on two brackets near the nose of the aircraft in such a position that:

(*a*) it is clear of the slipstream from the propellor or exhaust,

(*b*) both thermometers can be easily read over the whole temperature range by an observer inside the aircraft looking through a window,

(*c*) the bulbs are equidistant from, and at least 10 cm. clear of, the skin of the aircraft,

(*d*) the honeycomb grille nearest to the dry bulb faces forward.

In the mounting conforming to these requirements the instrument has a speed correction coefficient α_D of 2·1 (for temperature in degrees Fahrenheit and speed in knots). The method of installation is illustrated in Plate XXXIX. The two brackets should be made locally as their shape depends on the type of aircraft. Two holes are provided for bolting the instrument to the brackets, one through a locating block near the top and the other through the lower part of the water container. The locating block may be adjusted, after loosening a grub-screw, up or down the thermometer-protecting tubes through a distance of about 3 cm. if required.

As supplied the instrument is ready for mounting on the starboard side of the aircraft. If, however, it is necessary to mount it on the port side the thermometers and their protecting tubes may be turned through 180° if the grub-screw near the top and the hexagon locknut at the lower end are first loosened.

Replacement of a thermometer is easily made by unscrewing the milled head of the protecting tube and lifting out the spring and washer holding the thermometer in its mounting.

The operation of the instrument for the measurement of humidity is dealt with in Section 10.4.

10.3.4. Meteorological Office flat-plate electrical thermometers for aircraft

Description.—The general principles of electrical resistance thermometers and their application to meteorological purposes are discussed in Section 3.3 of Part I. That section also includes a full description of the Meteorological Office aircraft thermometer comprising the flat-plate electrical resistance element, Mark 1 (Stores Ref., Met. 2732), and the balanced-bridge temperature indicator, Mark 1B (Stores Ref., Met. 2828), since this equipment may also be used for surface observations. As the Mark 1B indicator is graduated for readings on the Fahrenheit scale, another model, the Mark 2B (Stores Ref., Met. 1525) with graduations on the Celsius scale covering the range + 40° to − 80°, is now used for aircraft observations. Each indicator is individually calibrated with its particular thermometer element and is provided with a correction card. If either part is renewed the whole must be re-calibrated.

The lag coefficient of the flat-plate thermometer is 8 sec. for an airspeed of 180 kt. at 900 mb. and it varies inversely as the square root of the airspeed and of the density.

Installation.—The thermometer element in its radiation shield is mounted near the leading edge on the underside of the starboard wing in the case of a single-engined aircraft or on the underside of the nose of the fuselage of a multi-engined aircraft. It must be orientated so that the flat surface of the element is parallel to the direction of air flow. In this mounting the flat-plate thermometer has a speed correction coefficient (for airspeed in knots and temperatures in °C.) of 0·95. The case of the radiation shield has 4B.A. clearance holes for bolting to the wing or fuselage and it is necessary to make a $\frac{5}{16}$ in. diameter hole for the connecting lead. This is connected to a three-way terminal block inside the wing or fuselage and from there suitable lengths of Tricel 4 cable are passed to the indicator control unit, the wires of the cable being connected to those of corresponding colours of the thermometer element lead.

The best position for the indicator depends on the type of aircraft. The essential point is that the user must be able to read the dial with his eye squarely in front of the indicator, and be able to operate the switches and the knob. The terminal block must also be readily accessible. An anti-vibration type of mounting must be used for the indicator. Confirmation that the leads are correctly connected may be obtained by noting that, with the indicator switch on, the pointer of the galvanometer should move in the same direction as the upper edge of the knob when that is turned.

Although the instrument is designed to take a considerable out-of-balance current through the galvanometer, this should be avoided as far as possible by making sure that the dial is set to the approximate air temperature before switching on.

10.3.5. Meteorological Office conical-head thermometer, Mark 1

Description.—This instrument (Stores Ref., Met. 1662) which was designed by Clark[116] and is shown in Plate XL, serves the same purpose as the flat-plate thermometer but is intended for use on high-speed aircraft. It differs from the flat-plate instrument in having improved aerodynamic properties in high-speed air flow, and in having a more rapid response to temperature changes. The principle of the design is that air flows along the slant surface of a cone and communicates its temperature to a polished aluminium frustum situated about one-third of the distance from the tip along the conical surface. The temperature of the frustum is measured by means of a platinum resistance element wound on a small former fitting inside the aluminium surface.

The construction is shown in Fig. 36. When the parts are fitted together they form a cone of semi-angle 10° and about 10 cm. in length. The tip of the cone (1) is made of hardened Monel metal and is highly polished. It has a threaded shank which passes through the former of the resistance element and screws into the main body of the cone, but it is insulated from the aluminium thermometer cover (2) by a Tufnol ring. The thermometer cover is electrolytically polished on the outside in order to reduce radiation errors and is anodized on the inside to insulate it electrically from the thermometer element. Fitting closely against the inside of the thermometer cover is the resistance element (3) consisting of a platinum wire wound in the grooves of a threaded aluminium former. The main body (4) of the

METEOROLOGICAL OFFICE AIRCRAFT ANEROID BAROMETER, MARK 2B

ROYAL AIR FORCE SENSITIVE ANEROID ALTIMETER, MARK 14

Front view of Altimeter

Plate XXXVIII

METEOROLOGICAL OFFICE AIRCRAFT PSYCHROMETER, MARK 6

INSTALLATION OF PSYCHROMETER, MARK 6, ON A BOSTON AIRCRAFT

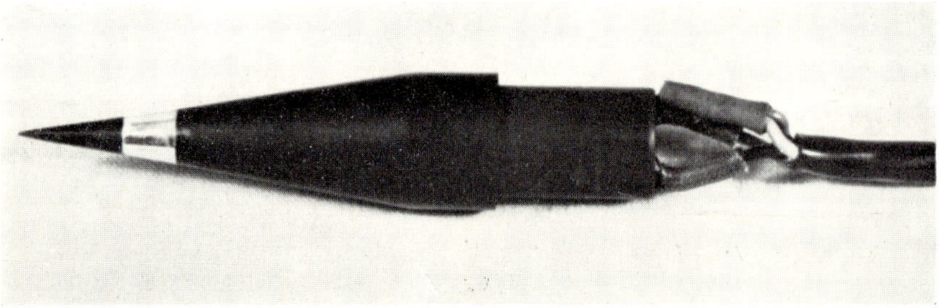

METEOROLOGICAL OFFICE CONICAL-HEAD THERMOMETER, MARK 1

MOUNTING OF CONICAL-HEAD THERMOMETER ON AN AIRCRAFT

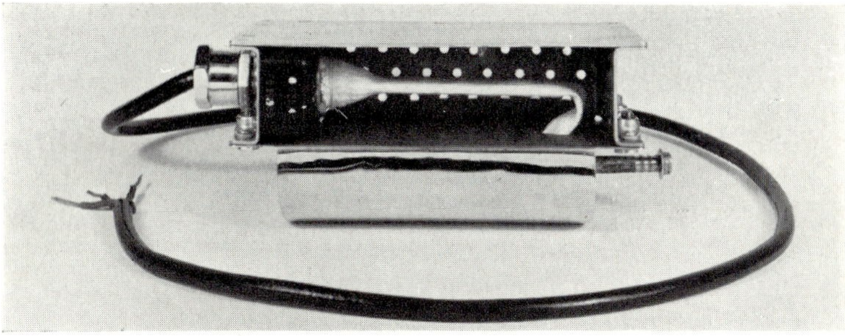

WET-BULB HOUSING OF METEOROLOGICAL OFFICE AIRCRAFT PSYCHROMETER, MARK 2

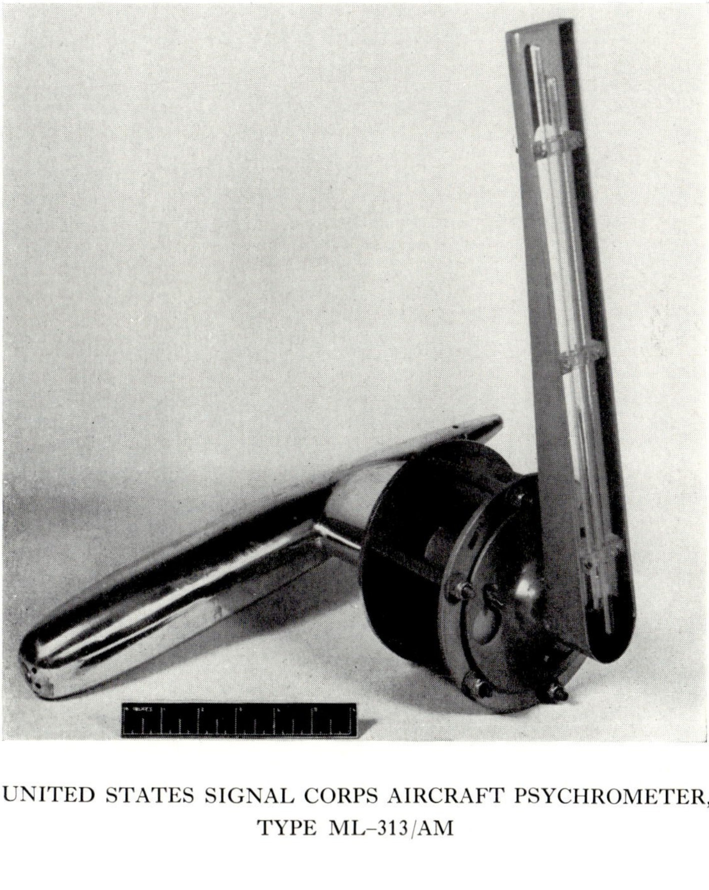

UNITED STATES SIGNAL CORPS AIRCRAFT PSYCHROMETER, TYPE ML–313/AM

instrument is made from Tufnol rod turned in the shape of the frustum of a cone with a cylinder at the larger end. It has a threaded hole at the other end into which the shank of the conical tip screws, and a recess in the cylinder to hold a ballast resistor (5). The latter is wound on a small spool and is wired in series with the resistance element; its purpose is to facilitate the adjustment of the instrument, during manufacture, to the specified resistance. Two terminal tags attached to the cylindrical end of the thermometer body provide the external connexions to the instrument.

The lag coefficient of the conical-head thermometer, defined as the time taken to respond to $1/e$ of a sudden change in air temperature, is about 2 sec. for an airspeed of 180 kt. at a level of 900 mb. It varies inversely as the square root of the airspeed and density.

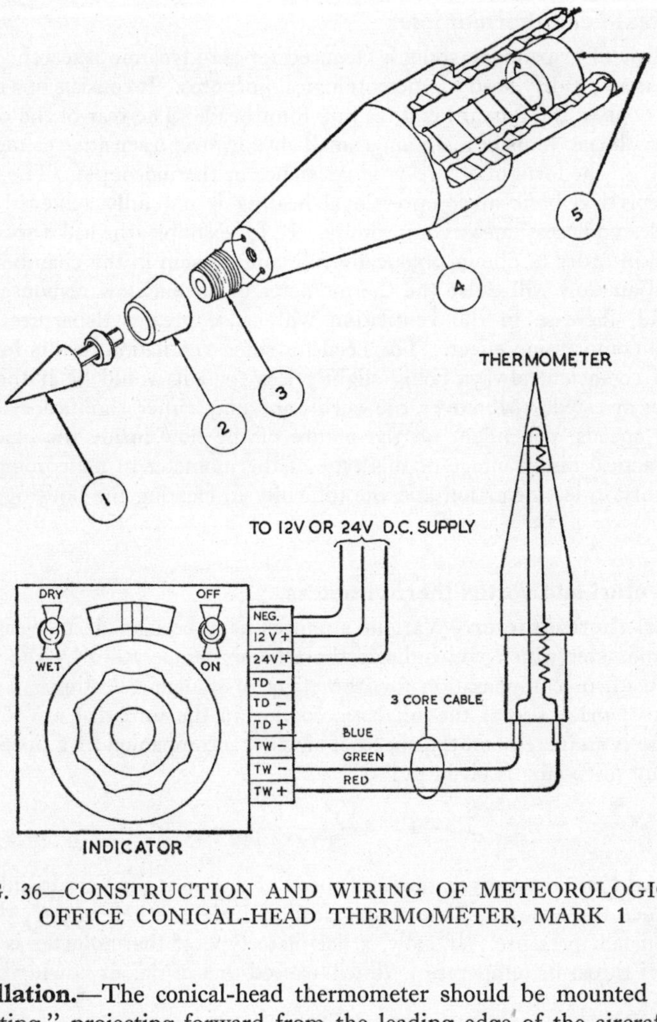

FIG. 36—CONSTRUCTION AND WIRING OF METEOROLOGICAL OFFICE CONICAL-HEAD THERMOMETER, MARK 1

Installation.—The conical-head thermometer should be mounted on a long metal " sting " projecting forward from the leading edge of the aircraft wing so that the thermometer is in undisturbed air at least 1 m. in front of the wing, as shown in Plate XL. The " sting " must be faired to the thermometer body so that there is a smooth join with no step or rough edge. It must also be hollow to allow connecting wires to be passed through to the aircraft. With this type of mounting the conical-head thermometer has a speed correction coefficient (for airspeed in knots and temperature in Celsius degrees) of 1·17.

The instrument is designed to be used with a balanced-bridge indicator, Mark 2B, and the connexions are similar to those for the flat-plate thermometer, except that as only two terminals are provided on the conical-head thermometer the green and blue wires from the indicator must be joined together at the thermometer end, and not at the indicator end, since the green wire acts as a compensating lead. As in the case of the flat-plate thermometer each conical-head instrument is calibrated with a particular indicator and if either part is removed the whole must be re-calibrated.

10.3.6. Total-head thermometer

This type of instrument, though favoured for aerodynamic research, is not, for various reasons, widely used for meteorological purposes. It consists of a cylindrical chamber, open at the forward end, as in a Pitot head. The rear of the chamber is provided with leak vents to maintain a small air flow over a sensitive element, which is generally in the form of a bare-wire resistance or thermocouple. The advantage of the theoretical basis of compressional heating is not fully achieved since the practical design necessitates a compromise. If, for example, the leak vents are made very small in order to obtain practically full compression in the chamber then the very small air flow will cause the thermometer to have a slow response. On the other hand, increase in the ventilation will cause greater departure from the theoretical compression effect. The need for some ventilation results in the speed correction coefficient always being slightly less than it would be if the air were completely arrested. Moreover the coefficient may suffer significant changes at very high speeds, dependent on the nature of the flow inside the chamber. A serious practical disadvantage of this type of thermometer in meteorological work is its liability to ice formation and the difficulty in clearing the housing once it is iced up.

10.3.7. Venturi and vortex thermometers

Venturi thermometer.—Various attempts have been made to design an aircraft thermometer which will indicate the true air temperature by the automatic elimination of, or compensation for, the effect of dynamic heating. Terada and Yamamoto[117] made use of the adiabatic cooling in the throat of a Venturi tube, where there is an increase in the speed of air flow accompanied by a pressure drop. The amount of cooling is given by:

$$T - T_1 = \frac{V_1^2 - V^2}{2JC_p},$$

where T and V are the temperature and speed of the air at the Venturi inlet, T_1 and V_1 the values at the throat, J the mechanical equivalent of heat and C_p the specific heat at constant pressure. If, now, a flat-plate type of thermometer is placed in the Venturi throat its temperature T_2 will exceed that of the air flowing over it by:

$$T_2 - T = \frac{\sigma^{\frac{1}{2}} V_1^2}{2JC_p},$$

where σ is the Prandtl number.

For complete compensation T_2 must equal T and therefore:

$$1 - \frac{V^2}{V_1^{\,2}} = \sigma^{\frac{1}{2}} = 0\cdot 85 \text{ (for laminar flow)} .$$

Continuity of air flow requires that $r^2V = r_1^{\,2}V_1$ where r and r_1 are the radii of the inlet and the throat of the Venturi tube respectively (density being assumed to remain constant). To achieve compensation, therefore, the dimensions required should be such that

$$(r_1/r)^4 = 0\cdot 15 .$$

In Terada and Yamamoto's instrument r_1 and r are 1·2 cm. and 2 cm. respectively and the thermometer is in the form of an enamelled nickel-wire resistance element flush with, but thermally insulated from, the wall of the Venturi throat, as indicated in Fig. 37. A flight test at a height of 4 km. showed that the temperature rise for an increase in speed from 140 to 220 kt. was about 0·7°C., which is about one quarter of the airspeed correction for a normally exposed flat-plate thermometer in these conditions. Evidently some modification of the dimensions of the Venturi tube was needed for complete compensation.

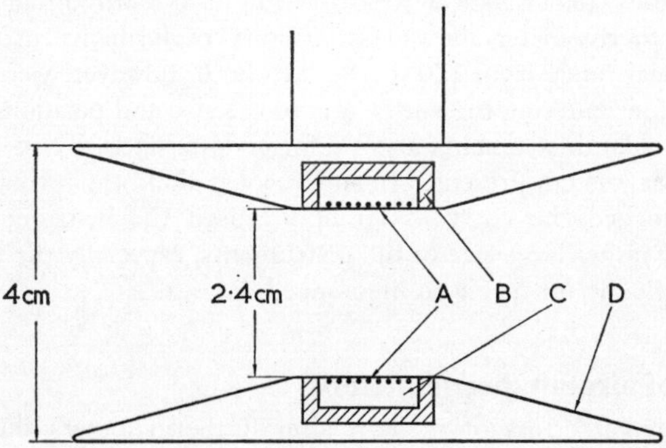

FIG. 37—VENTURI AIRCRAFT THERMOMETER
A, wire resistance element; B, ebonite; C, glass-wool; D, steel.

Vortex thermometer.—In this type of instrument use is made of the adiabatic cooling, consequent on the reduced pressure, in the centre of an air vortex. A housing designed for this purpose by Vonnegut[118] is illustrated in Fig. 38. It consists of a hollow cylinder with the thermometer element along its axis, the axis being transverse to the airstream. The air enters the cylinder tangentially and acquires rotary motion around the element as well as axial movement along the inside of the cylinder. The amount of cooling obtained at the centre of the vortex is controlled by a valve which regulates the air flow into the vortex. Flight tests showed that the housing reduces the effect of dynamic heating to less than 0·5°C., at least up to speeds of about 250 kt. and heights up to 8 km. In later designs of vortex thermometer the tangential inlet has been replaced by an axial inlet to the cylindrical housing, the axis of which is now parallel to the direction of flight. The vortex is produced by a spiral vane inside the cylinder.

In addition to the advantage of eliminating or greatly reducing the airspeed correction, vortex thermometers offer some prospect for the measurement of air temperature in clouds. The rotational velocities are probably sufficiently high to prevent cloud particles from depositing on the thermometer element and hence errors arising from evaporation or freezing of water on the element should be eliminated.

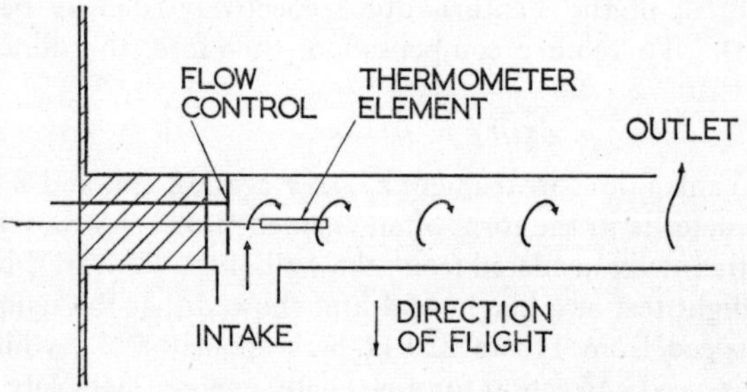

FIG. 38—VORTEX AIRCRAFT THERMOMETER

In tests by the Meteorological Research Flight at Farnborough in 1953 both types of vortex thermometer showed satisfactory performance in clear air up to 8 km. over a speed range from 120 to 180 kt. Both, however, were very sensitive to the amount of air entering the vortex and to the size and position of the thermometer element. The instruments also performed satisfactorily in cloud provided supercooled water was not present. In supercooled cloud the formation of ice near the intakes disturbed the entry of air and caused the instruments to become inefficient. The rather large size of the instruments, especially the tangential type, made them unsuitable for fitting to high-speed aircraft.

10.3.8. Errors of aircraft thermometers

The instrumental errors of modern aircraft thermometers that are specially designed for meteorological observations rarely exceed 0·5°C. even before calibration. With those that are individually calibrated, as is the case with the Meteorological Office instruments, instrumental errors may be as low as one or two tenths of a degree. The effect of self-heating in the electrical resistance thermometer of the Meteorological Office type, Mark 1, is, as explained in Section 3.3.4 of Part I, not significant at aircraft speeds.

Lag errors may be significant unless suitable precautions are taken to ensure that the aircraft does not pass too quickly through the level at which an observation is required. Thus with a rate of ascent of 6 m./sec. and a lapse rate of inversion of 10°C./km. a lag coefficient of 10 sec. will cause an error of 0·6°C. As the coefficient itself varies inversely with the mass rate of air flow ($\lambda \propto 1/(\rho\, v)^n$ where n may vary between 0·5 and 0·7), high-altitude observations may have large lag errors unless, as is usually the case, they are made in high-speed aircraft.

At the high ventilation rates to which aircraft themselves are subjected radiation errors are not likely to be important. In some types of thermometers they are minimized by the provision of radiation shields and by mounting the instruments on the under surface of the aircraft. In a slow-flying aircraft radiation errors may be significant if the attitude of the aircraft is such as to expose the thermometer element to direct radiation. The use of a radiation shield on a high-speed aircraft is objectionable on account of the drag that it introduces. In the Meteorological Office conical-head thermometer the radiation shield is dispensed with, the absorption of radiation being kept as low as possible by the highly polished surface covering the thermometer element. At an airspeed of 500 kt. such a thermometer has a radiation error of about 0·2°C. at 900 mb. increasing to about 1°C. at 100 mb.

Generally the largest errors in aircraft temperature observations arise from the speed correction coefficient. The difficulties associated with the determination of the coefficient from observations in flight lead to appreciable scatter in the values. In addition, with some types of thermometer element and mounting there are systematic changes of the coefficient with altitude. The accuracy of the coefficient is also affected by errors in the other instruments involved in the measurements, namely the airspeed indicator and the altimeter. Some information on the magnitude of the variations in speed correction coefficients for Meteorological Office flat-plate and conical-head thermometers has been given by Hinkel[119]. It is based on flights up to 12 km. at speeds from 160 to 420 kt. Over these ranges the conical-head thermometer showed practically no variation, and up to 10 km. the standard deviation of individual determinations of the speed correction coefficient was less than 3 per cent. of the mean value, but above 10 km. at high speeds the scatter nearly doubled. The flat-plate thermometer showed little variation with speed but did show an increase with height, amounting to about 15 per cent. of the coefficient, between the heights of 3 and 12 km. The standard deviation also increased, from about 3 per cent. at low levels to 7 per cent. above 10 km. Wind-tunnel experiments by Clark[120] suggest that the variations with altitude shown by the flat-plate thermometer are due in part to boundary layer separation and in part to the presence of temperature gradients on the thermometer surface. Differences shown in flight experiments, however, indicate that the results are also affected by the angle of attack of the thermometer element and by the angle of attack of the aircraft itself at high levels. It is by no means certain that the scatter in measurements of speed correction coefficient is due to the thermometers. As mentioned above, it could well be caused by errors of the airspeed indicator or altimeter.

As to the effect of the scatter on the accuracy of the temperature observations, it may be noted that for a standard deviation of 3 per cent. in the speed correction coefficient at heights up to 10 km. the error in the true air temperature is less than 1°C. at true airspeeds of less than 500 kt. Above 10 km. with a standard deviation of 6 per cent. the error is only less than 1°C. up to 300 kt. and is between 1° and 2°C. between 300 and 500 kt.

10.4. HUMIDITY MEASUREMENT FROM AIRCRAFT

10.4.1. General requirements

There are two main requirements for humidity measurements from meteorological aircraft. One is for a hygrometer that is suitable for routine measurements, to an accuracy of about 5 per cent., on reconnaissance flights at temperatures down to about − 30°F. The other requirement is for an instrument for measurements at high levels where temperature may fall below − 60°C. As in the case of radiosonde measurements, the chief difficulty in meeting these requirements, particularly the second one, is the very large range of water-vapour content in the atmosphere. Not only does the saturation pressure fall very rapidly as the temperature falls, e.g. at − 60°C. to 1/1300 of that at + 15°C., but humidities at high levels may be as low as 1 or 2 per cent., whereas near the ground values lower than 30 per cent. rarely occur.

The wide range to be covered is illustrated by considering a measurement in which the volume of the air sample is one cubic metre. At + 15°C. and 75 per cent. relative humidity all the water in the sample would provide a layer 1 cm. thick on an area of 10 cm.^{2} At − 60°C. and 1 per cent. relative humidity the layer would be only 10^{-5} cm. thick. Clearly, it is not easy to arrange for one type of instrument to measure, with comparable accuracy, these vastly differing quantities. In order to measure the small amounts of water vapour that may occur at high levels a large volume of air must be sampled. One possibility, therefore, of covering the wide range of measurement with one type of instrument is by providing a means of controlling the size of sample. Another possibility is to have a suitably wide adjustment of the sensitivity. It may be more convenient, however, to use separate instruments for the high-level and low-level conditions.

The fact that the rate of response of humidity-sensitive devices such as hair and gold-beater's skin decreases rapidly with decrease of temperature as well as with decrease in ventilation rate should not be as important in aircraft applications as in radio-sondes, since very high ventilation rates can be provided in aircraft and so offset, to some extent, the effect of temperature. A high ventilation rate is essential, in any case, for high-level measurements by a surface absorption or deposition method in order to pass large sampling volumes over the surface. The employment of high ventilation rates, however, necessitates the use of robust humidity elements and so such delicate materials as gold-beater's skin and hair are not generally used.

Methods, such as infra-red absorption technique in which the measurement is made through a large volume of air, have been tried on aircraft on an experimental basis, but for routine observations the methods in which a large quantity of air is forced on to or past a surface are used. They fall into two main types, one being the aircraft psychrometer, both liquid-in-glass and electrical forms, and the other the frost-point hygrometer. The latter is dealt with in detail in Section 10.5.

10.4.2. Aircraft psychrometers

Meteorological Office aircraft psychrometer, Mark 6.—This instrument has already been described in Section 10.3.3 in connexion with the measurement of air temperature. In order to obtain the maximum accuracy of reading of the wet-bulb depression the two thermometers in a psychrometer are matched so as to have, as far as possible, the same calibration errors and they should not differ by more than ± 0·2°F. This accuracy is necessary if relative humidity is to be obtained with an accuracy of ± 2·5 per cent. at temperatures down to about 20°F.

Installation of the psychrometer should include arrangements for the replenishment of water in the reservoir. A pipe is led from a tank inside the aircraft to the reservoir on the psychrometer and a priming pump is used to refill the reservoir periodically. Operational instructions, including the use of the instrument when the wet-bulb is ice-covered and the effect of lag, are given in the *Meteorological air observer's handbook*[112]. It is normally possible to continue to obtain ice-bulb readings down to temperatures of about 0°F.

Meteorological Office aircraft resistance psychrometer, Mark 2.—This instrument (Stores Ref., Met. 1743) comprises the electrical dry-bulb thermometer and balanced-bridge indicator, described in Section 10.3.4, together with a wet-bulb resistance element and housing, Mark 1A (Stores Ref., Met. 121). The wet-bulb element is of the same flat-plate type as the dry-bulb and only differs from it in having a stronger mounting at the base since it is not supported at the other end, whereas the dry-bulb is supported by its radiation shield. The housing for the wet-bulb element is shown in Plate XLI and it combines the functions of a radiation shield and a water reservoir. The radiation shield is in the form of a rectangular housing open at its forward end and pierced with ventilation holes at its rear end. A water reservoir of aerofoil shape is bolted to the underside of the housing, which is provided with a hole through which the wick sheathing the resistance element passes. A pipe connexion is fitted to one end of the reservoir so that water can be replenished during flight from a supply in the aircraft. If this facility is not needed the connexion should be closed with the screw provided. All outer surfaces of the housing and the reservoir are nickel-plated and polished.

In the installation of the equipment the wet-bulb housing should be mounted alongside the dry-bulb so that they are side by side when viewed from the front of the aircraft. The housing is provided with 4 B.A. clearance holes for bolting to the underside of the aircraft wing or fuselage and it should be orientated so that the perforated wall of the radiation shield is to the rear. The electrical connexions to the appropriate terminals on the balanced-bridge indicator are similar to those for the dry-bulb element and are shown in Figs. 52 and 53 of Part I, where information on the method of use and on accuracy are given. The speed correction coefficient for the wet-bulb element differs slightly from that of the dry-bulb and has the value of 1.1 for speed in knots and temperature in degrees Celsius.

American aircraft psychrometer.—The United States Signal Corps psychrometer, ML–313/AM, for meteorological observations from aircraft (shown in Plate XLI) is a liquid-in-glass type but it is not provided with a water reservoir. Instead, provision is made for dipping the wet bulb in water during flight. The thermometers are of right-angle form on a mounting which fits into an aluminium housing attached to the outside of the aircraft cabin. This allows the bulbs to be exposed to the air flow through the housing while the vertical stems with the scales are supported inside the cabin. The mounting is held firm by a bayonet type of joint which allows it, together with the thermometers, to be withdrawn from the housing into the cabin for dipping the wet bulb. With this arrangement the wet bulb can be kept in operation to lower temperatures than is practicable with the reservoir type of psychrometer. The housing for the bulb is streamline in shape with a double conical air passage leading from ventilation holes in the nose of the housing.

Lag and speed correction coefficients of the wet bulb.—It is shown in the *Meteorological air observer's handbook*[112] that the lag coefficient of a wet bulb is smaller than that of a dry bulb of the same form and heat capacity, and that the difference decreases with decrease of temperature. The dependence on the mass rate of air flow (ρV) is the same in the two cases. In practice the heat capacity of the wet bulb is greater than that of the dry bulb because of the wick and water or ice covering. Nevertheless, at temperatures near the freezing point the lag of the wet bulb is of the order of half that of the dry bulb and unless account is taken of this factor errors in the wet-bulb depression may be appreciable.

The effect of dynamic heating on the indications of the wet bulb is dependent on the dry-bulb heating effect and the ratio r between the adiabatic lapse rates for saturated and dry air appropriate to the pressure and temperature of the wet bulb. Thus if the rise in temperature of the latter when the wick is dry is $\alpha_W (V/100)^2$ the indicated wet-bulb depression will be too large by $\{\alpha_D - \alpha_W r (V/100)^2\}$, where α_W and α_D are the wet-bulb and dry-bulb speed correction coefficients. For the Mark 6 psychrometer the values of α_W and α_D do not differ appreciably. For the electrical-psychrometer wet bulb in its housing, Mark 1A, the value of α_W is 1·16 (for airspeed in knots and temperature in degrees Celsius), which is appreciably higher than the value of 0·95 for the dry bulb; this is probably due to the greater tendency for the wet-bulb housing to arrest the air.

10.5. FROST-POINT HYGROMETERS

10.5.1. General remarks

The development of the classical dew-point hygrometer of Regnault into an instrument for measuring from aircraft the very low water-vapour contents at high levels was successfully accomplished by Dobson, Brewer and Cwilong[121,122]. The success of their design, which depends on the deposition of frost from a jet of air on a cooled surface whose temperature can be controlled and measured, is due mainly to the fact that the surface can withstand air impinging on it at high speed and therefore a relatively large sample can be measured in a short time. Several models of the frost-point hygrometer have been developed for use in Meteorological Office aircraft, starting with a simple manually controlled instrument, followed by a similar type in a pressurized version and then by experimental designs, some still under development, in which a photo-electric cell is used to detect the frost deposit and automatic control of the cooling is provided.

An advantage of the dew-point and frost-point hygrometer is that it measures directly the temperature of saturation and since the depression of this below the air temperature does not vary rapidly with the latter the accuracy of measurement required to obtain relative humidity to the nearest five per cent. remains of a reasonable magnitude throughout the range of atmospheric temperature.

10.5.2. Meteorological Office frost-point hygrometer, Mark 2B

This instrument (Stores Ref., Met. 1104) is the current model for use in non-pressurized aircraft. Air from outside the aircraft is led, from a forward-facing scoop, through a jet across the black surface of a thimble which is cooled by a regulated flow of petrol, itself cooled by solid carbon dioxide. A deposit of frost is formed on the black surface and the true hoar frost-point is the mean of the temperatures of the thimble surface when (*a*) the amount of frost deposit is just decreasing and (*b*) when it is just increasing. The thimble temperature is measured with a platinum resistance thermometer and a similar thermometer is provided for measuring the air temperature, so that relative humidity may be derived. The instrument is illustrated in Plate XLII and Fig. 39. It comprises four main components, namely, the thimble, cooling system, illumination system and balanced-bridge temperature indicator.

The thimble.—This is a hollow cylinder, closed at the top, made of pure aluminium anodized and dyed black, with the top surface highly polished. The platinum resistance wire with which the thimble temperature is measured is wound in a screw-thread cut in the skirt of the thimble and is protected by a coat of bakelite

METEOROLOGICAL OFFICE FROST-POINT HYGROMETER, MARK 2B

Plate XLIII

METEOROLOGICAL OFFICE PRESSURIZED FROST-POINT HYGROMETER, MARK 3

varnish. Enamelled copper wire leads provide connexions to a balanced-bridge indicator, Mark 2C (Stores Ref., Met. 1602), which can be switched over to an air-temperature resistance thermometer as required. A small heater coil wound round a former surrounding the skirt of the thimble is provided for removing the deposit on the thimble should natural heating be insufficient. It is controlled by means of a press-button switch.

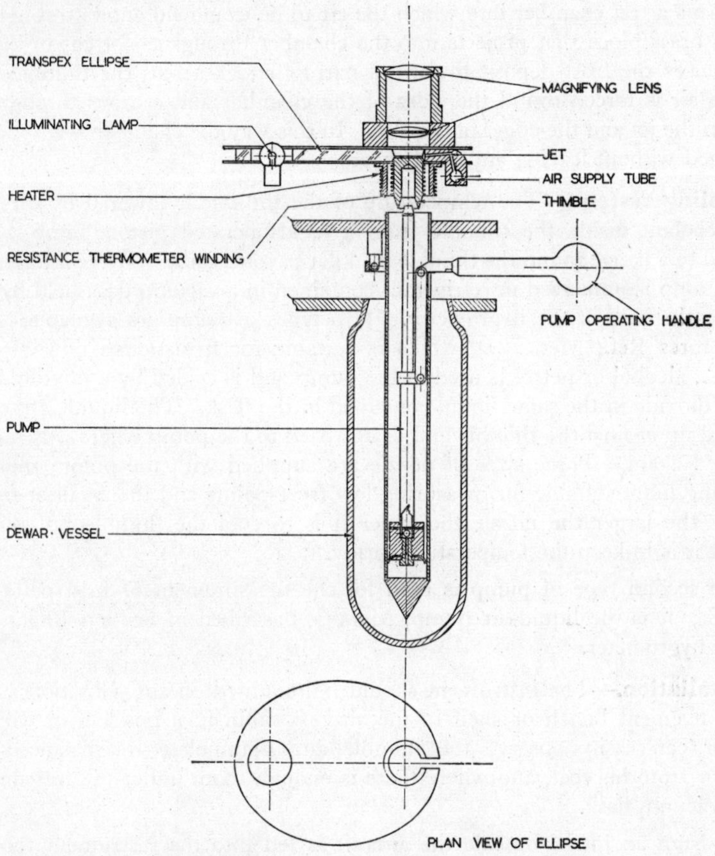

FIG. 39—CONSTRUCTION OF METEOROLOGICAL OFFICE FROST-POINT HYGROMETER

Illumination system.—This is a novel feature of the original design. The surface of the thimble is illuminated, at almost grazing incidence, by light from a 24 V. 6 W. lamp concentrated by an elliptical glass plate with holes centred at each focus. The lamp, the brightness of which is controlled by a rheostat, fits into one hole with its filament at the focus and the thimble top fits into a seating formed by the bevelled lower edge of the other hole. The upper edge of this hole is also bevelled, but in the opposite direction so as to refract light on to the thimble surface and at the same time reflect back into the ellipse light reflected upwards by the lower edge. As the outer edge of the glass ellipse is silvered all the light is internally reflected from the one focus to the other. This illumination over the black surface of the thimble provides maximum contrast of the ice crystals against their background and enables the very small crystals associated with low frost-points to be detected. The frost deposit is viewed through a wide-field, deep-focus lens system mounted on a metal plate which covers the upper surface of the ellipse; the lens system, which consists of a pair of doublets, has a magnification of about 6·5 and it is fitted with a rubber eye-guard.

The space bounded by the lower lens surface, the thimble top and the bevelled hole forms a test chamber into which the air to be examined enters through a fine jet in a brass block that projects into the chamber through a slot cut in the glass. This causes the frost deposit to form a narrow streak across the thimble surface and the air is forced round the sides of the chamber and escapes through a gap between the jet and the edges of the slot. In this way the chamber is automatically scavenged without leaving any dead spaces.

Cooling system.—The temperature of the thimble is lowered by directing a jet of coolant inside the thimble from a hand-operated piston pump which is attached to a flange round the thimble by a flat bayonet joint. The cylindrical body of the pump is immersed in refrigerant contained in a vacuum flask held by elastic cords to the base of the hygrometer. Two types of pump are available. In one type (Stores Ref., Met. 727), which is suitable for frost-points down to about − 70°C., alcohol or petrol is used in the pump and is cooled by a mixture of solid carbon dioxide in the same liquid contained in the flask. The liquid in the pump is forced up against the thimble and drains back to the pump where it is re-cooled and used again. Three sizes of nozzle are supplied with the pump, the largest size being more suitable for measuring low frost-points and the smallest for dew-points; the larger the nozzle the easier it is to cool the thimble but the more difficult it is to keep the temperature steady.

The second type of pump is used for the measurement of frost-point below − 70°C.; it is the liquid-air pump, Mark 3, described in Section 10.5.3 on the Mark 3 hygrometer.

Installation.—The instrument should be mounted on anti-vibration supports on a convenient bench or shelf in the aircraft cabin in a position in which the operator can easily observe the thimble and manipulate the balanced-bridge indicator from his seat, and where there is enough room under the instrument to fit the vacuum flask.

Air from an intake outside the aircraft is fed into the instrument through a clean metal tube connected to the screwed union provided for the purpose. To ensure an adequate flow through the hygrometer the air should be at a pressure of about 20 mb. above atmospheric and this is achieved by an intake in the shape of an open-ended tube a few inches outside the aircraft skin pointing forward, in a position not affected by the engine exhausts. It is essential that no rubber or non-metallic tubing should be used in the air supply pipe as it may alter the humidity of the air by absorbing or giving off water vapour.

The air supply tube should never be allowed to become wet inside. It is therefore necessary to fit a simple shutter, operated from inside the aircraft, for closing the intake in cloud and rain. It is desirable also, to include a three-way cock in the tube, with a connexion to another tube leading outside the aircraft, so that a continuous stream of air may be passed through the supply tube (but by-passing the hygrometer) in order to dry out any water accidentally admitted. The shutter should be kept closed during take-off and landing to exclude dust.

The installation and electrical connexions of the air-temperature element are similar to those of the dry-bulb element of the electrical resistance psychrometer, Mark 2, and the connexions of the thimble resistance thermometer are similar to those of the wet-bulb element but are incorporated in the instrument. The balanced-bridge indicator, Mark 2C, fitted to the hygrometer differs from the Mark 2B merely in the positioning of the terminals, switches and galvanometer for convenient manipulation in this application.

Preparation for use.—First, the most suitable nozzle should be fitted to the pump. In general the lower the frost-point to be measured the wider the nozzle required. A nozzle may be changed by extracting it with pliers and pushing another in its place.

The pump should then be filled with alcohol or petrol by pouring in at the top, taking care that no naked lights are in the vicinity. To fix the pump to the hygrometer the handle is unscrewed and the pump inserted through the large hole below the thimble. With the pump turned so that the screwed hole for the handle is in line with the slot in the base of the hygrometer, the three holes in the pump flange can be fitted over the three bolt heads under the base and by turning the pump slightly it is locked in position. The handle should then be replaced. As the upper part of the pump barrel is made of very thin metal to avoid conduction of heat, it should be handled carefully to avoid damage.

The cooling mixture is made by nearly filling the vacuum flask with powdered solid carbon dioxide and pouring alcohol or petrol in very slowly until it covers the solid material. The mixture should then be stirred to form a " slush ". Gloves must be worn when handling solid carbon dioxide. A flask full of the cooling mixture should last about two hours before all the carbon dioxide evaporates. Immediately before the hygrometer is to be used the flask is placed round the pump and pushed up until its neck fits into the hole in the base plate of the instrument, when it is fixed in place by the small plate and elastic cords.

Method of use.—The balanced-bridge is switched on and the change-over switch set for the thimble thermometer. The flow of air through the instrument is started and the lamp switched on. Full brilliance should normally be used unless working in semi-darkness, when reduced light will be preferable. Then, while observing the thimble through the lens the pump handle is moved up and down, so cooling the thimble with a jet of coolant. As soon as a deposit of dew or frost is observed the bridge is balanced and, while watching the galvanometer pointer, pumping is continued at such a rate as to keep the thermometer steady. After an interval of time varying from a few seconds for temperatures above 0°C. to several minutes at − 75°C. the thimble is inspected again. If the deposit has increased the bridge is set again to a slightly higher temperature (or vice versa), held steady for an appropriate time, and the thimble again inspected. In this way two temperatures will be found at which the deposit slowly increases and decreases respectively, and the mean of these is taken as the dew-point or frost-point. The two temperatures will differ by only a fraction of a degree at dew-points above 0°C. but the difference increases to about 5°C. at − 75°C. At such low frost-points the rate of growth and evaporation of the deposit is very slow and the measurement may take at least five minutes, during which time the humidity of the air may vary appreciably and so very accurate results could not be expected. In order that a very small change in the amount of the deposit may be clearly visible the deposit should be as thin as possible, and if necessary the thimble should be warmed until only a slight deposit remains.

The procedure of noting the temperature when the first deposit is seen and when the last trace evaporates does not give accurate results; neither does that of observing the thimble continuously while operating the pump.

At temperatures between 0° and − 20°C. the deposit may be either ice or supercooled water, and the thimble temperature giving a constant deposit will be slightly different in the two cases. They are distinguished by the grey appearance of the water and white appearance in the case of the ice. If a water deposit is used the resulting dew-point may be converted to frost-point by adding corrections as follows:

		degrees Celsius				
Dew-point	..	0	− 5	− 10	− 15	− 20
Correction	..	0	0·5	1·0	1·5	2·0

At temperatures between − 20° and − 40°C. the deposit may also be supercooled water on occasions, but it is then preferable to cool the thimble quickly in order to freeze the water and then warm it up until the deposit is thin enough to allow the frost-point to be determined in the usual way.

Testing and maintenance.—The calibration of the balanced-bridge indicator should be checked at three-monthly intervals. This is done by first removing the cover plate and ellipse and inserting an accurate mercury thermometer with its bulb resting on the thimble and keeping it in position by loosely replacing the cover plate. When the reading is steady the bridge, switched to the thimble thermometer circuit, is balanced and read. After applying the correction from the correction card the reading should agree with that of the mercury thermometer to within 0·3°C. If greater errors are found the whole instrument, including the air-temperature element, should be returned to the Instrument Branch of the Meteorological Office.

If the top of the thimble becomes dirty it should be cleaned with a soft cloth. When the lamp has to be replaced the height of the lamp holder should be adjusted, if necessary, by loosening the fixing bolt so that the filament of the lamp is in the plane of the ellipse. When replacing the ellipse care should be taken that it is the right side up, so that it fits closely round the thimble. The pump must be kept clean, and should preferably be removed and emptied after each flight and kept in a warm dry place.

10.5.3. Meteorological Office pressurized frost-point hygrometer, Mark 3

Description.—This instrument (Stores Ref., Met. 1652) which is illustrated in Plate XLIII and Fig. 40 is intended for use in aircraft with pressurized cabins and it differs from the Mark 2B hygrometer in having the thimble observation chamber sealed off from the cabin air and in not having the balanced-bridge indicator attached. The pump and the coolant are on the other side of the seal, at cabin pressure. This ensures that the pump continues to work at high altitudes and also facilitates replenishment of liquid air or nitrogen which is used as the coolant and is contained in an easily detachable vacuum flask. No heater is provided for the thimble since heat conduction from the body of the hygrometer is sufficient for removing the frost deposit when necessary.

The body of the instrument consists of three main parts. The upper part carries the lens system and is clamped, with a rubber-ring seal, to the middle

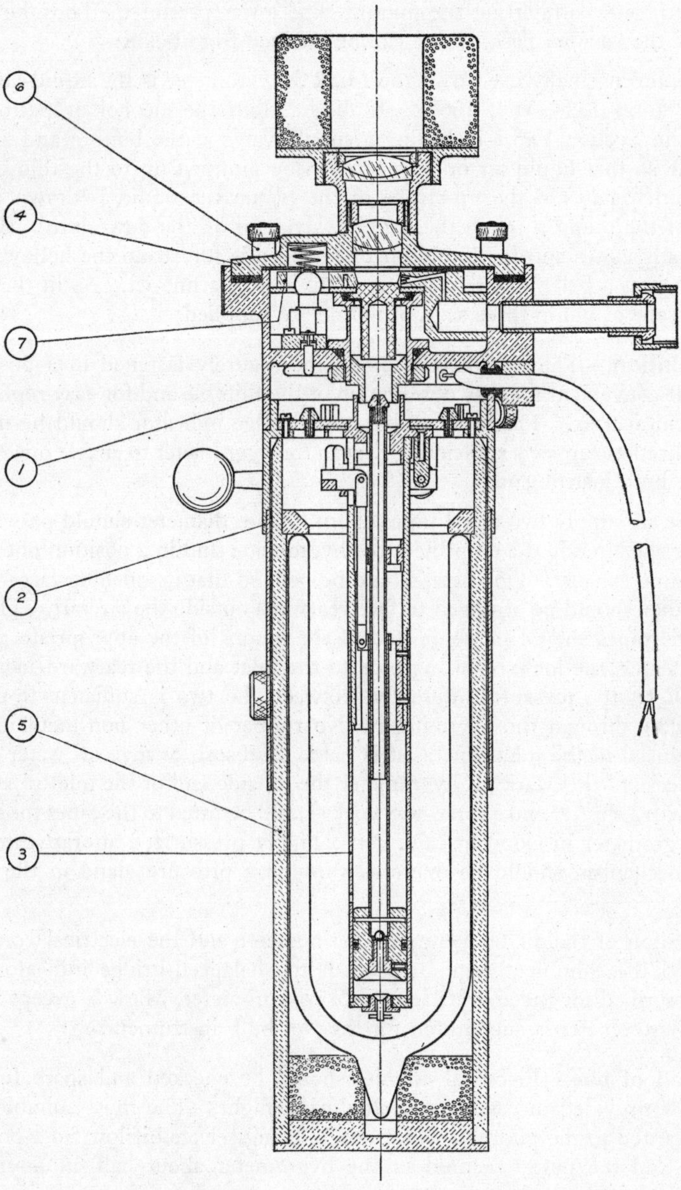

FIG. 40—CONSTRUCTION OF METEOROLOGICAL OFFICE PRESSURIZED FROST-POINT HYGROMETER, MARK 3

1, pressure chamber; 2, pump housing; 3, liquid-air pump; 4, cover and lens system; 5, vacuum-flask container; 6, special nut; 7, glass ellipse.

section. This contains the thimble, with its resistance-thermometer winding, the illuminator system and the air supply and exit ducts; it also has a flat, bayonet type of fitting for supporting the pump. The lower part of the body houses the pump and the vacuum flask, which can be inserted from below.

The pump normally used with the Mark 3 hygrometer is the liquid-air pump, Mark 3 (Stores Ref., Met. 1660). It differs from the alcohol or petrol pump described in Section 10.5.2 in having a supply valve at the bottom and an outlet at the side so that liquid air or nitrogen can be pumped up to the thimble from the vacuum flask. On the up-stroke of the piston the liquid is drawn into the cylinder of the pump through the supply valve and on the down-stroke the valve automatically closes and the liquid air or nitrogen is forced up the hollow duct in the piston past a ball valve to issue from the top in a fine jet. As in the case of the other type of pump three sizes of nozzle are supplied.

Installation.—The hygrometer should be securely fastened in a position in the aircraft convenient for easy observation of the thimble and for easy replacement of the vacuum flask. The separate balanced-bridge indicator should be mounted on anti-vibration supports sufficiently near to the hygrometer to enable one observer to operate both instruments.

For the air supply two metal pipes about 0·5 in. diameter should pass through the skin of the aircraft at a point near the hygrometer and in a position not affected by the engine exhaust. One pipe should be bent so that its opening faces forward and the other should be arranged to face rearward outside the aircraft. The other ends of the pipes should be connected to the unions of the appropriate pipes on the hygrometer, the forward-facing pipe to the inlet and the rearward-facing pipe to the outlet; the pressure differential between the two is sufficient to ensure a good air flow through the hygrometer. No rubber or other non-metallic tubing should be used in the inlet pipe as it is liable to absorb or give off water vapour. As in the case of the Mark 2B hygrometer the outside end of the inlet pipe should be fitted with a shutter and a three-way cock should be fitted to the pipes for shutting off the hygrometer in cloud or rain. In a highly pressurized aircraft the shutter opening mechanism should be operated through a pressure gland in the aircraft skin.

Installation of the air-temperature thermometer and the electrical connexions of this and the thimble thermometer with the balanced-bridge indicator are as already described for the aircraft resistance psychrometer, Mark 2, except that the thimble thermometer is substituted for the wet-bulb instrument.

Method of use.—Electrical circuits should be checked and spare fuses and lamp bulbs provided in case of a fault during flight. The most suitable nozzle should be fitted to the pump (a wider nozzle being selected if low frost-points are expected) and the pump secured in the hygrometer about half an hour before commencing a flight.

The vacuum flask should be partially filled with liquid nitrogen or air and then slowly inserted into the hygrometer housing from below, the clamping screw then being tightened. A few strokes of the pump at this stage will help it to cool more quickly. After about five or ten minutes the flask is removed, topped up with coolant to within about an inch from the top, and replaced. Thick gloves must be worn during these operations as liquid nitrogen and air can burn if care is not taken. The coolant may be replenished during flight if a reserve flask of it is carried, but this practice is not recommended and as far as practicable sufficient coolant should be carried in the hygrometer to cover the required measurements. It is therefore advisable to top up the coolant as near take-off time as possible, to minimize wastage due to evaporation.

Soon after the aircraft is airborne and before it is pressurized, air should be allowed to flow through the hygrometer pipes, so blowing out any dust that may have accumulated, and the thimble should be cleaned. The top of the hygrometer should then be replaced and bolted down tightly and evenly on the rubber washer. With the balanced-bridge and lamp switched on, the hygrometer is now ready for use and the subsequent operations of the pump and observation of the frost deposit are similar to those described for the Mark 2B hygrometer.

Recognition of the frost deposit at low temperatures is often difficult. At temperatures less than about 10°C. above the frost-point, frost often forms on the edges of the thimble. This " background " icing should be ignored since it results from minute leaks in the hygrometer pressurization seals. It adds to the difficulty of observing as it tends to destroy the contrast between the jet area and the rest of the thimble surface.

It is emphasized that the thimble temperature may need to be held steady for up to a minute before frost can be detected with certainty. The temptation to watch the thimble continuously while operating the pump should be resisted.

Testing and maintenance.—The calibration of the hygrometer should be checked regularly and this can be most conveniently done when inserting the coolant before each flight. After the flask of coolant is first fitted into the hygrometer and the top of the instrument is removed the pump should be operated until a thick white deposit of frost is formed on the thimble (e.g. at about − 10°C.). (The growth of the deposit can be accelerated by breathing on the thimble.) The thimble is then allowed to warm up and its temperature noted on the indicator when the deposit begins to change from frost to water. This reading should be within 0·25° of 0°C. on the scale. The balanced-bridge indicator and air-temperature thermometer should also be checked occasionally.

The pump must be kept clean and dry; it should be removed after each flight and dried out. Occasionally it will be found necessary to dismantle the pump and renew the silastomer washer.

Method of extending the range.—At very low temperatures the frost deposit becomes increasingly difficult to recognize and at temperatures below − 85°C. the deposit may be in the form of glassy ice which is very difficult to observe without special optical arrangements involving the use of polarized light. A method of increasing the range of the hygrometer by compressing the air to a high pressure before reading its frost-point has been described by Goldsmith[123]. The installation in a jet aircraft is shown in Fig. 41. As an alternative to the outside air supply to the Mark 3 hygrometer in the pressure cabin, air from the engine compressor may be fed to the instrument through a control valve when the outside supply is shut off. Inlet and outlet pressures are measured and if the control valve is adjusted to produce surface pressure in the hygrometer the value of the frost-point obtained can be conveniently converted to the equivalent value at the aircraft height by means of a suitable thermodynamic chart such as a tephigram. By this method the practical limit of usefulness of the hygrometer can be extended by 20°C. or more. There is also the advantage that air in the hygrometer is at a higher pressure than air in the cabin and so any leaks that occur are outwards, whereas when air at outside pressure is used any leakage is inwards and, at very low frost-points, may cause a large error.

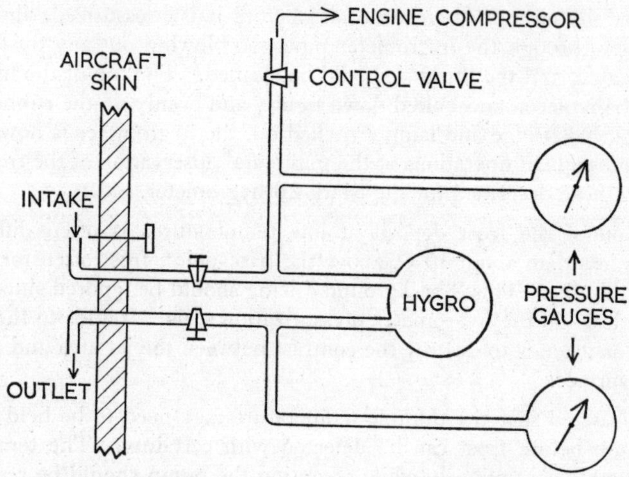

FIG. 41—USE OF COMPRESSED AIR TO INCREASE THE RANGE OF A FROST-POINT HYGROMETER

10.5.4. Errors of frost-point hygrometers

The accuracy obtainable by the frost-point method is largely determined by the accuracy with which the thimble temperature can be controlled. In the laboratory it is quite practicable to measure a frost-point of — 50°C. to within 0·5°C. since the temperatures at which the deposit increases and decreases do not differ by more than about 2°C. With care, and if time is not restricted, this accuracy can be obtained in an aircraft. At 500 mb. it represents an accuracy of about 5 per cent. in relative humidity.

Laboratory tests, reported by Brewer *et alii*[122] with air having a frost-point of — 78·5°C. (obtained by passing dry air through a copper spiral immersed in solid carbon dioxide and acetone) showed agreement of measurements by three observers to within 1°C. of that temperature.

A systematic error giving frost-points that are too low arises from a small temperature gradient that exists between the thimble surface and the thermometer winding. This may be of the order of 1°C.

10.5.5. Automatic frost-point hygrometers

Photo-electric indication of deposit.—Over the past 25 years there have been several designs of dew-point hygrometers in which visual observations of the dew deposit have been obviated by the use of a photo-electric method of detection. This application of photocells is, in fact, almost essential if the operation of the hygrometer is to be fully automatic. An example of such a hygrometer, though not for use in aircraft, is that of Thornthwaite and Owen[124].

In the development of automatic frost-point hygrometers for aircraft the first step was taken by Dobson and his co-workers[121,122], who adapted their instrument for the photo-electric detection of the frost deposit by replacing the visual optical system by the photocell arrangement shown in Fig. 42. The surface of the thimble is highly polished and when there is no deposit on it most of the light from the lamp is reflected into the black absorbing box and very little reaches the photocell. A small deposit of frost on the thimble, by scattering the light, causes a large increase in the light reaching the cell. The current passing through the photocell is amplified and read on a micro-ammeter. Thus the reading is a measure of the amount of deposit on the thimble. In use, the thimble temperature is lowered until a deposit is obtained; then by careful pumping of the coolant the temperature is adjusted until the deposit is neither growing nor evaporating, as shown by a steady indication on the micro-ammeter. The temperature of the thimble is then that of the frost-point.

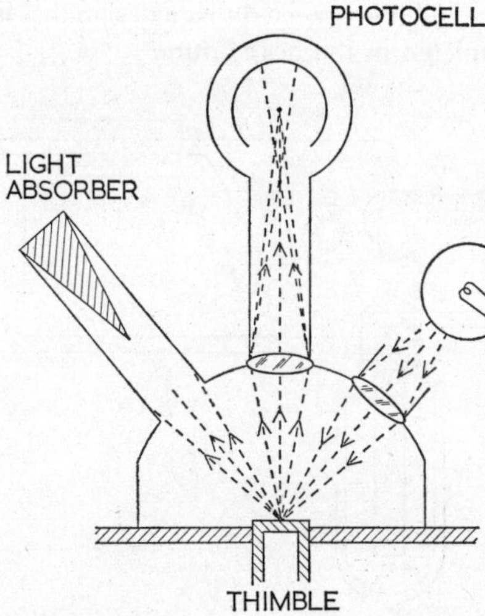

FIG. 42—PHOTO-ELECTRIC DETECTION OF FROST-POINT

Automatic hygrometers.—Although in principle it is only a small step to make the photo-electric instrument fully automatic in operation, by making the output from the photocell control the cooling of the thimble, the development of an automatic instrument sufficiently reliable for routine use in aircraft presents a number of difficulties. These are associated mainly with the enormous range of vapour pressures that has to be covered, the need for a reasonable speed of response and stability in operation, and restriction in size and weight.

The experimental designs that have been tried out in recent years have certain essential features in common; these are illustrated in Fig. 43. The thimble surface S is continuously cooled through a conducting rod dipping into coolant in a vacuum flask but its temperature is controlled by passing current through a heater coil H. Two photocells are used, one (P) to detect the light scattered by the deposit on the thimble and the other (Q) receiving light direct from the lamp. Their outputs are combined differentially in order to compensate for variations in lamp brightness and in the sensitivity of the photocells (which may arise from voltage variations or from fatigue). The differential output from the photocells is fed to a servo-amplifier which controls the current supplied to the heater coil. In one American design of hygrometer the thimble heating is provided by radio-frequency induction controlled by the servo-unit. With correct adjustment of the circuit the thimble is warmed if the frost deposit is large, while if the deposit is small the thimble cools. If the air passing over the thimble has a constant water content, equilibrium is quickly reached, with the thimble at the frost-point maintaining a constant deposit of ice. The temperature of the thimble may be measured by a thermocouple or resistance thermometer connected to an indicator or recorder. The servo control unit may take a variety of forms, such as a magnetic amplifier, an a.c. amplifier and light-beam chopper, or a phase-controlled thyratron. It is difficult to achieve stability of the control system over the very wide range of operation that is normally desired and some improvement in this respect may be obtained by using a circuit operating on the rate of change of error signal or by reducing thermal lag in the thimble. Experimental instruments have been operated successfully in the laboratory and it is expected that a design that is quite satisfactory for aircraft use will be completed in the near future.

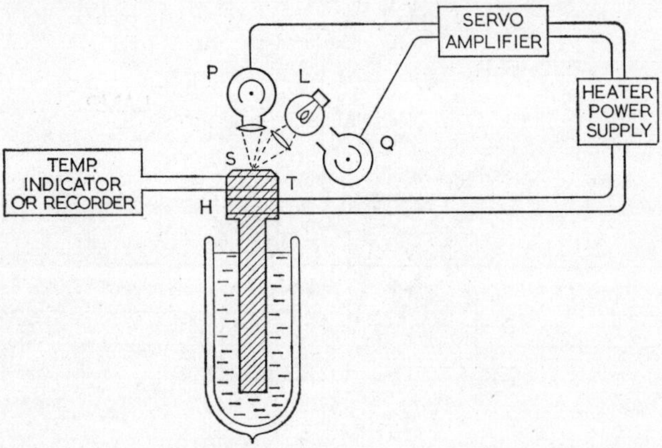

FIG. 43—ESSENTIAL FEATURES OF AN AUTOMATIC FROST-POINT HYGROMETER

10.6. INFRA-RED ABSORPTION HYGROMETERS

There have been several projects, principally in the U.S.A., to develop for use in aircraft a method of measuring absolute humidity based on the absorption by water vapour of infra-red radiation at certain wavelengths. The method consists essentially in the measurement of the attenuation by water vapour of a beam of radiation of a suitable wavelength passing through a column of air of known length. It has the great advantage that the measurement is made on the air itself and not through the action of some intermediary material such as hygroscopic film or a condensation or evaporation surface. The speed of response, therefore, is limited only by the speed of the indicating instrument.

An instrument developed for the United States Weather Bureau and tried out in an aircraft has been described by Wood, Foskett and Boster[125], and is illustrated in Fig. 44. For the source of radiation a tungsten lamp S is used, in front of which is a set of eight glass filters mounted as sectors of a wheel F which is rotated at 15 turns per second. Four of the filters transmit radiation in a narrow band centred near $1 \cdot 37\ \mu$, i.e. one of the water-vapour absorption bands, and these are alternated with four filters transmitting at about 1·24 μ where the absorption is inappreciable. The two bands are therefore interposed in the beam at a frequency of 60 c./s. After traversing the sample of air the beam is received by a lead-sulphide photocell L, the output of which is amplified to give a signal consisting of two 60 c./s. components with a phase difference of 180°. When the radiation flux through each set of filters is such that the two components are equal the combined signal is zero. If, however, water vapour is present in the sample the beam from the 1·37 μ filters will be attenuated and the lack of balance between the two components of the photocell will give an error signal. This is used to control a servo-mechanism which changes the voltage and therefore the temperature of the tungsten lamp until balance is restored. Thus the water-vapour content may be measured in terms of the lamp temperature; measurement is made, in fact, of the energy of the lamp emission at balance by means of a selenium photocell.

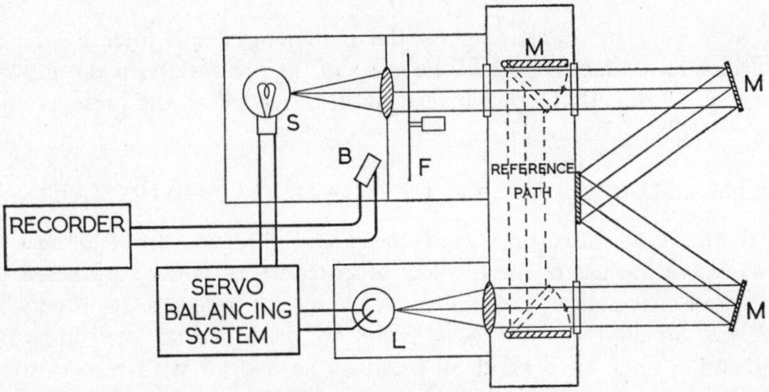

FIG. 44—UNITED STATES WEATHER BUREAU INFRA-RED ABSORPTION HYGROMETER

S, radiation source; F, filter sector wheel; M, mirrors; L, lead sulphide cell; B, selenium cell for measuring energy of source.

In adapting the equipment for aircraft tests a folded light path, using three mirrors M, was introduced in order to save space and, in addition, as a check on calibration drift, a reference path enclosing a sample of air of known humidity was provided. This reference path could be inserted in the system by turning two mirrors, as shown in Fig. 44. At concentrations of water vapour of less than 1 gm./m.^{3} the hygrometer can detect variations of less than 25 mgm./m.^{3}, i.e. less than 30 parts in a million; at − 40°C. this would represent a variation in relative humidity of 20 per cent.

An improved design of infra-red absorption hygrometer has been described by Wood[126]. It is similar in principle to the United States Weather Bureau instrument but incorporates some features which are reported to result in a better performance. Instead of the glass filters it uses germanium interference filters, for absorption at the wavelength of 2·60 μ and for the reference measurement at 2·45 μ. The balancing of the two components of the photocell output is done not by controlling the voltage of the light source but by moving a glass wedge in the beam, the movement being made by a transducer operated by the servo-mechanism. The wedge is made of glass that has different transmission coefficients for the two wavelengths. A precision potentiometer coupled to the wedge mechanism converts the movement into an electrical quantity for recording on a suitable meter. The optical path of the equipment is enclosed by a tube about 30 cm. long and 5 cm. diameter through which the sample can be passed or, as a check on calibration, air which has been dried in an attached desiccator. At concentrations of 0·2 gm./m.^{3} or less the hygrometer can detect variations of 1 mgm./m.^{3} of water vapour, or about one part per million by volume.

A difficulty in connexion with the application of the infra-red absorption hygrometer to upper air measurements arises from the fact that although the absorption is mainly dependent on the amount of precipitable water vapour traversed it also depends on pressure, and to a lesser extent on temperature. Wood obtained calibration curves for his equipment with samples at pressures of 1, $\frac{1}{2}$, $\frac{1}{4}$ and $\frac{1}{8}$ atmospheres and derived a relationship between A the energy absorbed, W the amount of water vapour, and P and p the total and partial pressures, of the form:

$$A = C\,W^{0\cdot 5}(P + p)^{0\cdot 23},$$

where C is a constant depending on the path length and filter characteristics. Other workers have found values of the index of the pressure term varying between 0·25 and 0·5. Evidently further investigation is needed on the pressure effect.

10.7. MEASUREMENT OF WIND FROM AIRCRAFT

The methods of measuring wind speed and direction from aircraft do not normally involve the use of meteorological instruments as they are based on well established air navigation procedures. These are referred to briefly in the *Meteorological air observer's handbook*[112] and are described in detail in manuals of air navigation. They all depend on locating the aircraft with respect to a fixed point on the earth's surface and, if the measurements are not to be restricted to clear or nearly clear air below the aircraft, radio or radar navigational aids must be used.

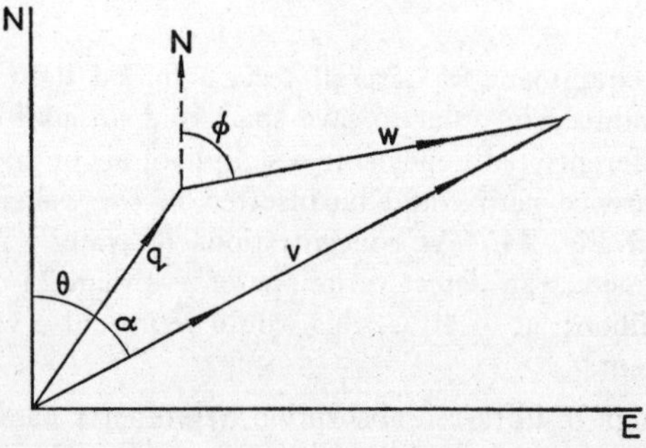

FIG. 45—VECTOR DIAGRAM OF THE RELATION BETWEEN GROUND SPEED v, AIR SPEED q AND WIND SPEED w

The wind measurement requires the determination of four quantities, namely, the heading of the aircraft obtained from compass readings, the ground speed and the angle of drift, both obtained from the navigational aid, and the true airspeed which is derived by standard methods from the indicated airspeed, the atmospheric pressure given by the altimeter and the indicated air temperature. From Fig. 45 it will be seen that the wind is the vector difference between the airspeed vector, given by the heading θ and the true airspeed q, and the ground speed vector derived from the heading θ, the ground speed v and the angle of drift α. If there are no errors in these measurements the wind speed w and direction φ are given by:

$$w \cos \varphi = v \cos (\theta + \alpha) - q \cos \theta$$
$$w \sin \varphi = v \sin (\theta + \alpha) - q \sin \theta\,.$$

The accuracy of such measurements using a Doppler navigational system has been investigated by Murgatroyd and Helliwell[127]. They found that with careful calibration not only of the electronic equipment but also of the instruments used in deriving the true airspeed vector the vector wind can be determined to within 5 kt., but that without these calibrations the error may exceed 10 kt.

CHAPTER 11

METHODS OF MEASUREMENT IN CLOUD PHYSICS

11.1. GENERAL

The study of the physics of cloud and precipitation, to which much effort has been devoted in recent years, has entailed the development of a number of new and specialized instruments. They have been developed mainly for research purposes and only a few have become sufficiently well established for use in routine observations for synoptic or climatological purposes. For this reason and because further development continues rapidly it is proposed to give only brief descriptions of some of the instruments that are in use at the present time.

We may distinguish between two main classes of equipment for measurements in cloud physics, depending on whether the measurements are made from the ground or from the air. The first category includes, for example, the optical methods of determining the height and movement of clouds which are described in Part I and the application of ground radar techniques to the detection and measurement of phenomena associated with cloud and precipitation. In the other class the airborne equipment that is involved is generally of the type that is carried in aircraft rather than by balloon. One of the few examples of balloon-borne instruments used for cloud-physics investigations is the alti-electrograph designed for detecting the electric fields of thunder clouds (Section 1.3.1). The obvious advantage of using aircraft rather than balloons for such investigations is that they allow almost complete control, up to the height limitation of the aircraft, in taking the instruments to the regions where the measurements are required.

The chief measurements in cloud-physics investigations with aircraft, in addition to those of temperature and humidity, are of the liquid water content and the size distribution of particles in clouds and of precipitation elements in and below clouds. A fundamental difficulty of such measurements is that of ensuring that the samples examined are representative of the natural conditions under investigation and that these conditions are not appreciably disturbed by the instruments or their mountings. This difficulty is particularly important in measurements of size distribution.

Because of the short times taken by modern aircraft to traverse clouds it is highly desirable, if not essential, for the instruments used to be capable of automatic recording and to have very small lag coefficients (of the order of a second or less). Automatic recording may be arranged either by photographing the dials of indicating instruments or by designing the instruments to give electrical outputs which can be recorded by a multi-channel galvanometer.

11.2. INSTRUMENTS FOR MEASURING WATER CONTENT

11.2.1. General considerations

The amount of free water in a cloud is generally of the order of 1 gm./m.^{3} but may be as high as 10 gm./m.^{3} or as low as 0·05 gm./m.^{3} The water may exist as cloud droplets of a few microns diameter, as raindrops up to several millimetres in diameter or as ice crystals, and the temperatures may be such that the water droplets are supercooled. It would be difficult to design one instrument to measure the water in all of these forms and to cover the wide range of possible variation of the quantity with an accuracy of, say, 10 per cent. There are, therefore, a number of designs in use, each having limited applications. Most of them involve exposing a collecting surface or opening to the air flow and measuring the water caught in a known time. The volume of air sampled per unit time is obtained from the true airspeed and the area of the surface, but to obtain the water content the efficiency of the collector must also be known. This depends on the shape and dimensions of the collector. No surface gives 100 per cent. efficiency of collection for all drop sizes since very small drops are carried round the surface by the air-stream. Theoretical derivations of collection efficiencies of surfaces of well defined shape are available but it is generally necessary to obtain empirical values by calibration.

Very few methods give satisfactory results when the water is in the form of ice crystals or a mixture of ice and liquid. It should be possible to obtain measurements in these conditions by using a heated duct to evaporate all the ice and water and then measure the dew-point but, so far, practical difficulties have hindered the development of a satisfactory instrument on this principle. One of the difficulties is that if a reasonable sampling rate is to be maintained an inconveniently large amount of heat is required.

Two examples of successful methods for use when all the water is in the liquid state are described briefly in the following section. One depends on the measurement of electrical resistance of a material wetted by the water and the other on the rate of ice formation from supercooled droplets.

11.2.2. Electrical resistance method

An instrument using the electrical resistance measurement has been described by Warner and Newnham[128]. It consists essentially of a paper strip which moves past a narrow slit exposed to the airstream where it is wetted by the cloud droplets. After a short delay to allow the paper to absorb the droplets the electrical resistance between two contacts resting on the paper is measured and recorded on a recording milliammeter. The paper is of the type used in electrochemical recording; the speed with which it moves past the slit can be varied between 5 and 60 cm./min. and a counter measures the amount of paper used. By a suitable choice of paper speeds, water contents ranging from 0·05 to 5 gm./m.^{3} can be measured. The slit is located at the end of a narrow housing projecting from the leading edge of the aircraft wing and it can be closed when necessary by a motor-driven shutter.

The instrument is not suitable for use in moderate or heavy rain or in the presence of supercooled water droplets as the slit becomes covered with water. It is subject to the usual sampling errors by failing to collect all of the smallest droplets present in clouds, the average collection efficiency in a typical cloud being about 90 per cent. at aircraft speed. In calibrating the instrument with droplets sprayed into a wind tunnel it is assumed that the resistance of the paper tape depends only on the amount of water collected per unit area, which for a given liquid water content is a linear function of the ratio of the airspeed to the paper speed. The instrument is considered by its designers to be reliable to $\pm$ 20 per cent. for cloud temperatures of 10°C. There is a temperature coefficient of resistance of the paper of the order of — 2·5 per cent. per °C. Although the absolute indications are not of high accuracy the instrument does afford a convenient means of recording relative values of water content and the variations extending over a hundred metres or so.

11.2.3. Rotating-disc icing meter

For the measurement of the amount of supercooled water in clouds the rate of accretion of ice on slowly rotating cylinders exposed to the airstream has been used. The amount of icing is a function of the collection efficiency of a cylinder, and therefore of its dimensions, the airspeed and the liquid water content of the air. Originally the amount of accumulated ice was measured directly by withdrawing the cylinders from the airstream and weighing or measuring the thickness. A much more convenient arrangement is to use a thin rotating disc continuously exposed to the airstream and to measure continuously by an electrical method the thickness of ice accumulated on the edge of the disc. This is done by means of a feeler lightly pressing on the ice and operating a transducer in an a.c. bridge circuit, the unbalanced output of which produces a meter reading proportional to the thickness of the ice and therefore to the supercooled water content and the airspeed. A scraper behind the feeler removes the ice and so the accumulation during each revolution is measured. The upper limit of the range of such an instrument is about 2 gm./m.^{3}

At higher values of water content the latent heat produced by the change from water to ice eventually raises the temperature of the disc to a point where an appreciable amount of water impinging on it flows off instead of freezing. This limitation has been largely overcome in an instrument used by the Meteorological Research Flight and described by Day[129]. The disc in this instrument, a general view of which is shown in Plate XLIV, is hollow and is cooled by liquid nitrogen fed into it through a hollow driving shaft. The rate of supply of the coolant can be adjusted to maintain the disc at a given temperature. When this is very low the instrument is also capable of measuring water contents at cloud temperatures above freezing point. The instrument has been used to measure, continuously, water contents up to about 4·0 gm./m.^{3} and the accuracy is probably better than $\pm$ 10 per cent. in conditions normally encountered.

11.3. MEASUREMENT OF PARTICLE SIZE DISTRIBUTION

11.3.1. General

The solid and liquid particles that occur in clouds vary in diameter from a few microns in the case of cloud droplets to a few millimetres in the case of raindrops and even to centimetres for some hailstones. It is not to be expected, therefore, that any one instrument can be made to measure the sizes and concentration of particles over such a wide range and so the methods developed vary with the types of the particles. Most of the methods depend on the collection and subsequent examination of the particles but optical methods depending on scattering or attenuation of light by the particles have also been developed. The sampling of large particles is made difficult by their splashing or breaking on impact and quantitative measurements of the size distribution of hail, snow and larger raindrops are not yet on a satisfactory basis.

11.3.2. Instruments for cloud-droplet sampling

Cloud droplets usually occur in concentrations of a few hundred per cubic centimetre and range in size from a few microns to about 100 microns. They are

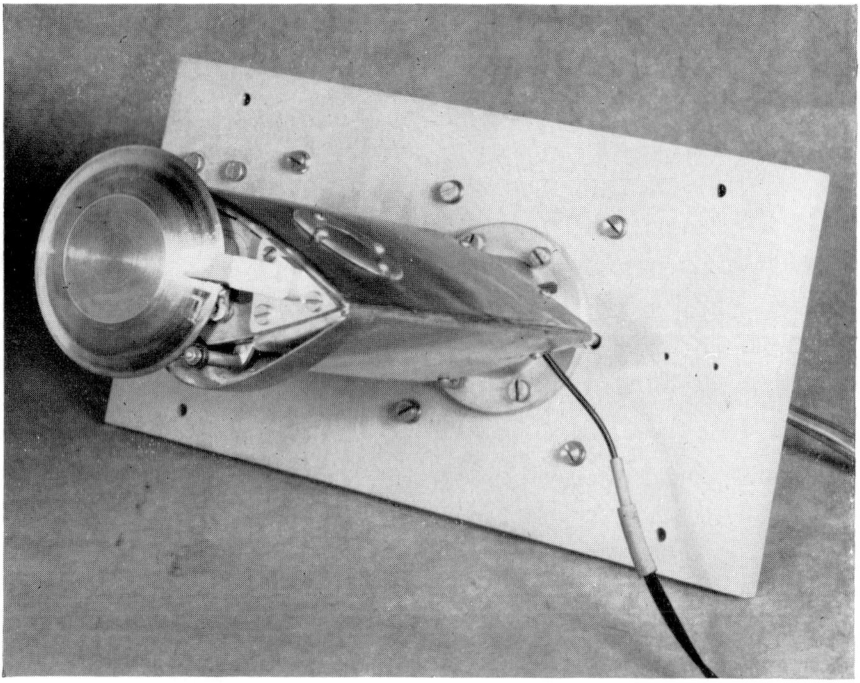

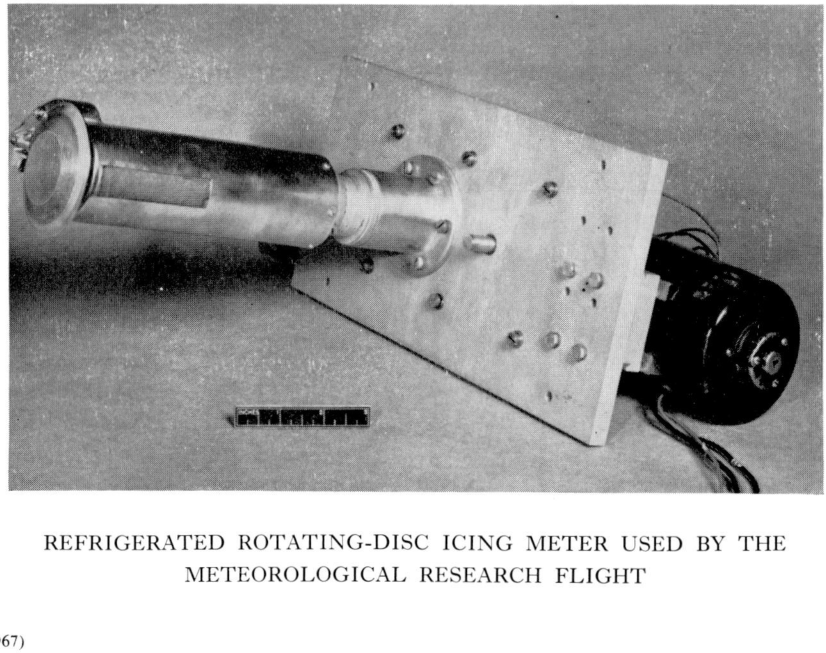

REFRIGERATED ROTATING-DISC ICING METER USED BY THE METEOROLOGICAL RESEARCH FLIGHT

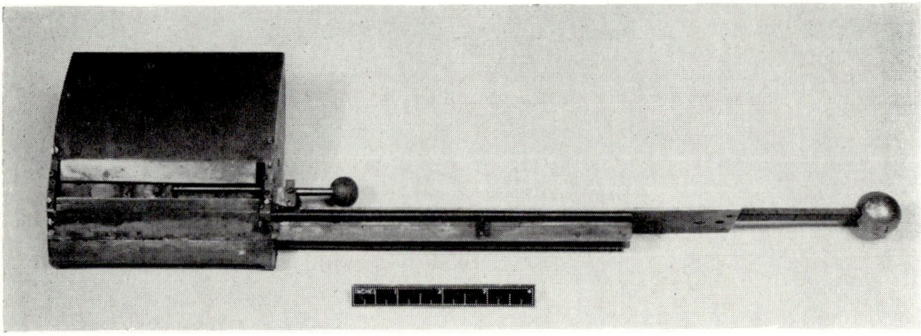

(*a*)

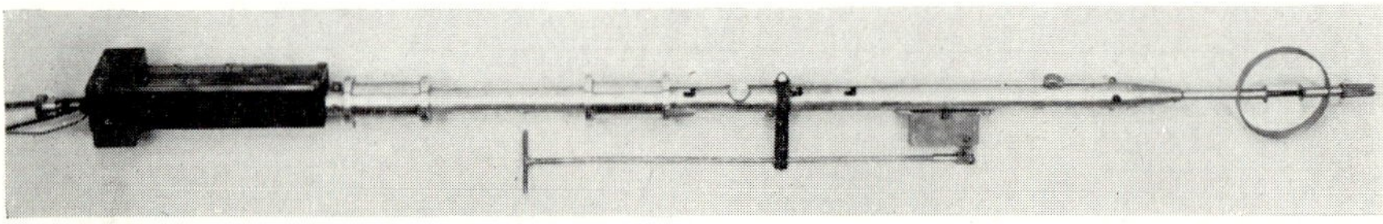

(*b*)

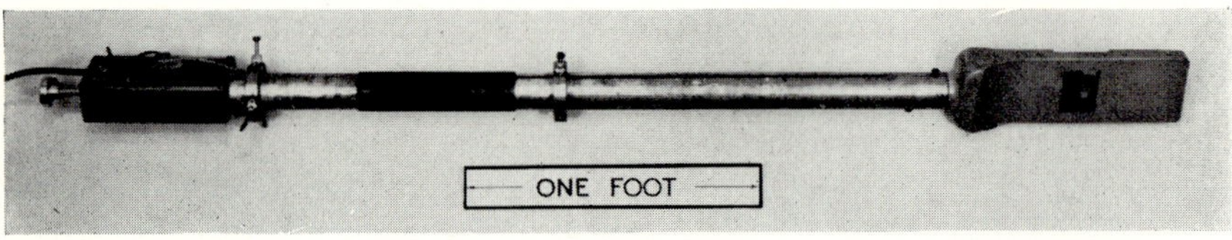

(*c*)

IMPACTORS FOR SAMPLING CLOUD DROPLETS, (*a*) AND (*b*), AND FOR SAMPLING CHLORIDE PARTICLES (*c*)

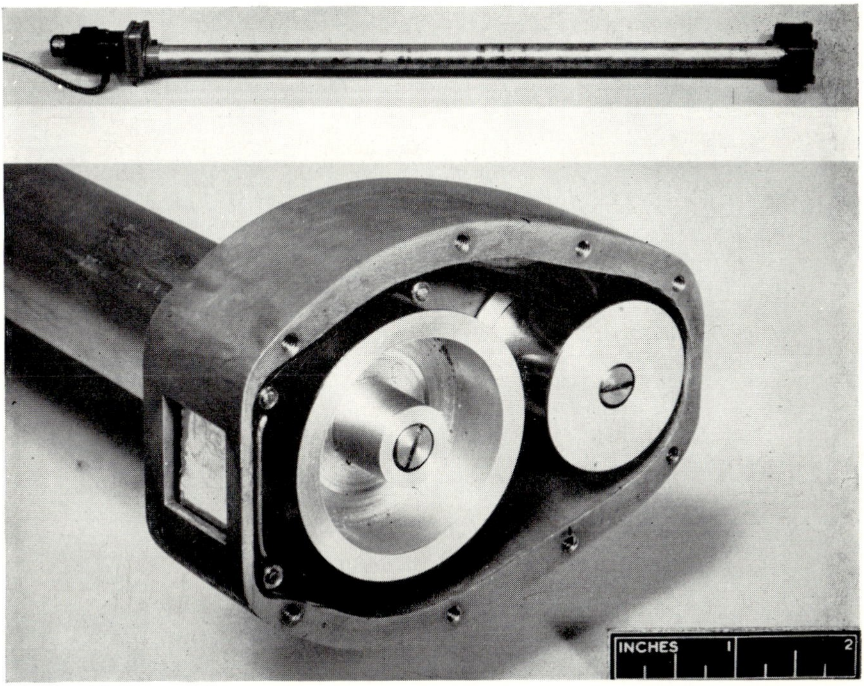

ALUMINIUM TAPE RAINDROP IMPACTOR

The lower photograph shows the exposed head.

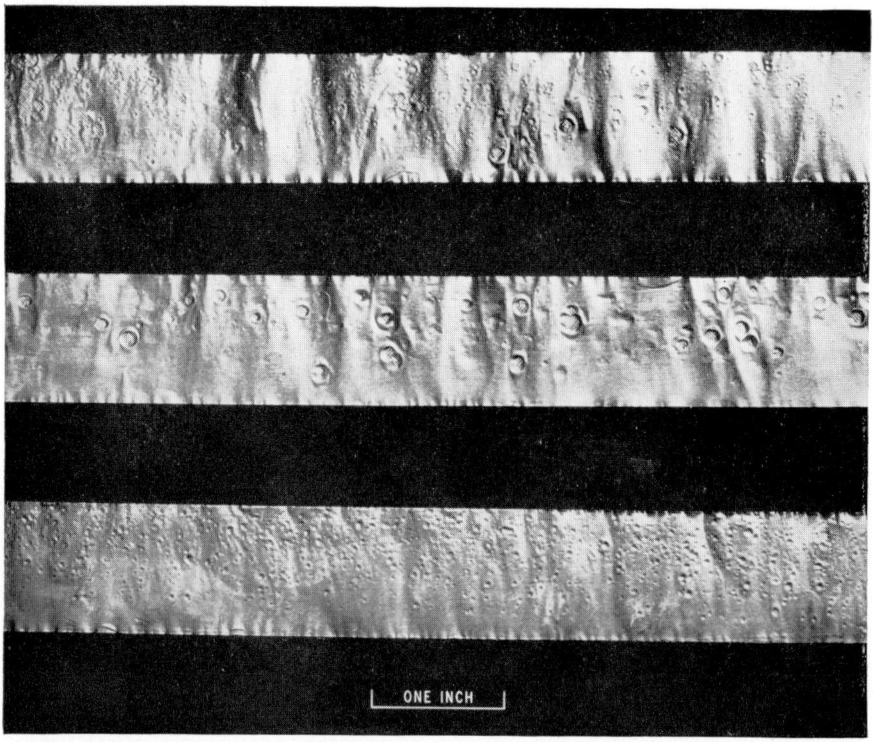

EXAMPLES OF PRECIPITATION IMPACTS ON ALUMINIUM TAPE

Top: mixed rain and snow. Centre: rain. Bottom: drizzle.

generally sampled by exposing collecting slides or rods to the airstream for intervals of a few hundredths of a second, the collecting surfaces being coated with a suitable material. Soot or magnesium oxide is commonly used and the droplets, on impact, make small craters in the surface, the size of the craters being proportional to the size of the droplets. This technique has the advantage that the surface can be measured or microphotographed at leisure after the end of a flight. It is, however, difficult to use at airspeeds above about 150 kt. and droplets larger than about 100 μ in diameter and also ice crystals shatter the surface.

An alternative method is to coat the slide with oil of a suitable viscosity so that the droplets penetrate the surface of the film and are captured in the oil. In this case, however, the examination or photography of the sample must be done quickly to avoid coalescence or partial evaporation of the droplets. Ice crystals can also be caught in the oil but unless photographed quickly will appear as large droplets.

Two methods of exposing the slide have been used in the Meteorological Research Flight. One, which is shown in Plate XLV (*a*), employs a simple shutter in front of the slide while in the other arrangement, shown in Plate XLV (*b*) and described by Durbin[130], the slide is shot across a gap or slot, the exposure in both cases being about 1/75 sec. The sampling head is at the end of the movable arm which can be exposed to the airstream from the cabin of the aircraft. The arm carries a trigger for operating the shutter or slide movement. When the slides are examined under a microscope or in a microphotograph the counting of the droplets in various size ranges is facilitated by using a scale marked with circles of diameters limiting the ranges.

If the exposure time is known and the collection efficiency of the sampling head is nearly 100 per cent. the method also provides a measurement of the water content. The sampling head is designed to minimize the effects of splashing. In the cloud-droplet sampler devised by Squires and Gillespie[131] glass rods, 3 mm. in diameter, are used instead of flat slides. The rods are coated with a layer of soot and are exposed for about a thirtieth of a second, thereby sampling a volume of air of about 50 cm.^{3}

11.3.3. Instruments for raindrop and ice-crystal sampling

For very small raindrops, of the order of a few hundred microns in diameter, an impactor using a sooted plate similar to that used for cloud droplets is suitable. For larger drops a sooted gauze is more satisfactory and at the Meteorological Research Flight it was found that a silver-plated gauze with a mesh of about 30 strands per centimetre is suitable. A raindrop does not splash in striking the gauze but passes through it and in so doing removes a small quantity of soot. A visible impression of the drop is left on the gauze and its size is related to that of the drop and to the speed of impact. The method is also suitable for obtaining impressions of small snow-flakes.

An improvement on this technique that is particularly suitable for measurements of large drops at high aircraft speeds has been described by Garrod[132]. In the course of investigations at the Royal Aircraft Establishment at Farnborough it was found that a raindrop striking thin aluminium foil in front of a wire-gauze screen at 100 kt. or more leaves a circular indentation and an inner circular mark formed by impressions of the gauze behind the foil. The inner circle has a diameter proportional to the true drop size. An impactor for continuous recording designed on this principle is shown in Plate XLVI. The sampling head consists of an aerofoil-shaped housing containing two spools, one holding about 6 m. of aluminium tape 0·005 cm. thick and 2·5 cm. wide and the other being driven by an electric motor and drawing the tape at a rate of about 7 cm./sec. past an aperture about 1 cm. wide in the front of the housing. Behind the aperture and the tape moving across it is a wire-gauze screen firmly soldered to a brass plate. Typical records obtained in different types of precipitation are reproduced in Plate XLVII. The instrument is suitable for measurements of ice particles as well as raindrops.

Another method which has been used, though not very successfully, for sampling ice crystals from aircraft is that of capturing the crystals on a slide newly coated with a quick-drying lacquer such as Formvar. When the lacquer dries replicas of the crystals remain visible. The method, which has been described in detail by Schaefer[133], is in frequent use for ice-crystal sampling in cloud chambers but the difficulty in applying it to aircraft measurements is in slowing down the speed of the air sufficiently to avoid fracturing the crystals.

11.4. MEASUREMENT OF ATMOSPHERIC NUCLEI

Atmospheric nuclei play an important part in the production of cloud and precipitation and information on their types, sizes and distribution (especially with altitude) is therefore fundamental to the study of the physics of clouds. The two main classes are the condensation nuclei and the freezing nuclei. Condensation nuclei are essential for the formation of cloud droplets from water vapour in saturated or slightly supersaturated air, while freezing nuclei are believed to play an important part in the Bergeron precipitation mechanism by inducing supercooled cloud droplets to freeze and so provide the ice crystals that are necessary for that mechanism.

11.4.1. Condensation nuclei

The concentration of condensation nuclei in the air is usually determined by means of an Aitken nucleus counter or a more modern version of the instrument. The original Aitken counter consists of a small chamber which is maintained at 100 per cent. relative humidity and in which the sample of air is subjected to sudden adiabatic expansion. The cooling is accompanied by a high degree of supersaturation and causes condensation on the nuclei present, which then fall as droplets on to a counting stage.

For the investigation of nucleus content in the free atmosphere it is much more convenient to use an instrument of the photo-electric type developed by Pollak and Nolan[134]. An instrument of this type is used in the Meteorological Research Flight and is shown in Plate XLVIII. The expansion chamber is in the form of a metal tube with glass ends and the wall of the tube has a porous lining which is kept moist. A parallel beam from a light at one end of the chamber passes through to the other end where it is received on a photo-electric cell. The installation includes a pump, an air filter and a pipe connecting the instrument with a suitable air intake fitted outside the aircraft.

The instrument is operated by first allowing air from the intake to flow through the chamber sufficiently long to clear it of all stale air. The air in the chamber is then compressed by pumping in filtered air until the pressure is higher than the ambient pressure in the ratio of about 1·22. After allowing the sample in the chamber to become saturated it is suddenly expanded by releasing the excess pressure and the extinction coefficient of the fog so formed is measured by comparing the photocell output with the value before the expansion. Spurious effects due to condensation on the glass plates are avoided by the use of electrically heated glass. The instrument is calibrated by comparison with an Aitken type of counter.

It may well be questioned whether all the nuclei counted in these instruments play any significant part in natural condensation processes. The degree of supersaturation produced by the expansion in the counter is of the order of 100 per cent. whereas condensation in the atmosphere takes place at supersaturations which rarely exceed 0·1 per cent. The counter detects all condensation nuclei with radii exceeding 10^{-6} cm. and the concentrations range from a few hundred per cubic centimetre over the sea to hundreds of thousands per cubic centimetre near the ground in industrial areas. On the other hand there are usually only a few hundred droplets per cubic centimetre in clouds. Evidently, therefore, the Aitken nuclei are far more numerous than true condensation nuclei except over the sea, where they may play some part in natural condensation. It is, in fact, almost certain that the nuclei that are effective in the natural condensation process are relatively large hygroscopic salt particles with radii greater than 10^{-4} cm.

Various techniques are in use for sampling these large salt nuclei, but most of them involve the use of some form of impactor. The method adopted in the Meteorological Research Flight is the chemical method known as the Liesegang ring technique, described by Vittori[135], and the impactor used is shown in Plate XLV (*c*). The air to be sampled is projected by the impactor on to a surface coated with a thin layer of gelatin impregnated with a soluble salt that will give a suitable reaction with the radical in the nuclei to be measured. For example, if chloride nuclei are to be sampled a soluble silver salt such as silver nitrate is used as the reagent. The chloride particles react with this to form a deposit of insoluble silver chloride and subsequent exposure of the deposit to bright sunlight reduces the silver chloride to metallic silver. The sample can then be microphotographed if necessary.

11.4.2. Freezing nuclei

Although the precise nature and origin of freezing nuclei remain in doubt the importance of these nuclei in the production of precipitation is generally accepted and has led to the development of a number of methods of measuring their concentration. They depend on the production, in a cold chamber, of a supercooled cloud in which nuclei present in the sample of air under test can induce freezing at a given temperature. The numbers of crystals formed at different temperatures are then taken to be the numbers of freezing nuclei active at these temperatures. Unfortunately the different methods that have been devised give discordant results, apparently depending on the technique adopted for producing the supercooled cloud. If this is done by gradually cooling the air and so keeping the supersaturation comparable with that occurring in the atmosphere many drops may fall out before freezing takes place. Also there is some loss of nuclei through condensation on the walls of the chamber. It seems probable that more consistent results can be obtained by cooling the air rapidly by expansion, but as in the case of condensation nuclei, there remains the doubt whether the measurements are representative of the natural cooling process.

An example of the slow cooling method is the cold cloud chamber apparatus designed by Smith and Heffernan[136] for measurements in aircraft. A similar installation is in use in the Meteorological Research Flight and the chief features are illustrated in Fig. 46. The chamber is a cylindrical copper vessel C of about 70 l. capacity which is cooled by circulating cold alcohol through pipes P coiled round the chamber, the pipes also being connected to a spiral immersed in solid carbon dioxide. The chamber is well lagged with insulating material and is covered with an insulating lid into which is fitted an illuminating system L projecting a parallel beam of light obliquely across the chamber and a rubber aperture which allows the beam to be viewed at right-angles. Two electrical thermometer elements T are provided for measuring the temperature of the air sample which is injected into the chamber after this has been thoroughly flushed out. Steam is introduced by electrically heating distilled water in the vacuum flask F. The cloud of water droplets has a dull foggy appearance in the light beam but any ice crystals present glisten brightly against the blackened walls of the chamber. The latter is cooled slowly to a temperature of about — 43°C. and the numbers of ice crystals appearing at different temperatures is estimated. Obviously the method does not lend itself to rapid sampling and the estimation of the numbers of crystals is subjective.

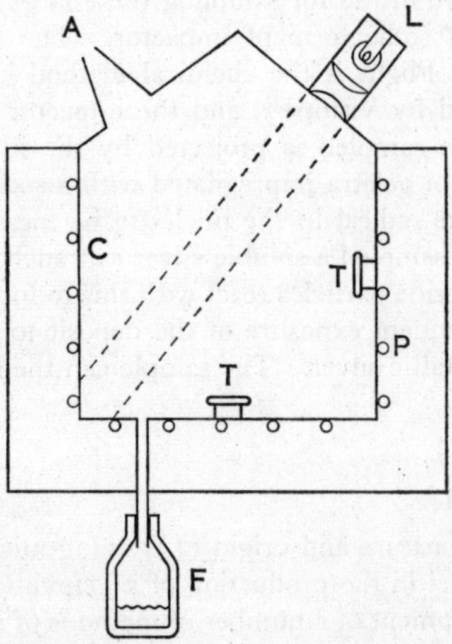

FIG. 46—CLOUD CHAMBER FOR OBSERVATIONS OF FREEZING NUCLEI (SMITH AND HEFFERNAN[136])

These disadvantages are, to a large extent, overcome by the expansion type of chamber, an example of which has been described by Warner[137]. This instrument has a chamber of 10 l. capacity which is cooled to — 12°C. At the bottom of the chamber there is a dish of sugar solution which supercools. After a sample of air is introduced the pressure inside the chamber is increased above the ambient pressure. Then the excess pressure is suddenly released and a supercooled fog is produced in which freezing nuclei may become active. Any ice crystals which are thereby formed fall into the supercooled sugar solution in which they grow rapidly and are easily counted. The temperature of the fog may be varied by varying the initial overpressure in the chamber.

Very few freezing nuclei are active at temperatures above — 20°C. but at — 30°C. numbers between 10 and 10^4 per litre have been reported.

11.5. APPLICATION OF RADAR TO CLOUD AND PRECIPITATION STUDY

11.5.1. General principles

The application of radar technique to the study of cloud and precipitation is based on the fact that the liquid and solid particles associated with these phenomena, i.e. the cloud droplets, raindrops, ice crystals, hailstones and snow-flakes, scatter microwave radiation and so produce radar echoes. Radar may therefore be used for such purposes as the determination of the vertical and horizontal extent of precipitation occurring over a wide area, the observation of the development and movement of belts of precipitation and of thunderstorms and the investigation of the relationship between echo intensity and rate of precipitation. The measurements of the radar echoes may include their range and azimuth from the observing station and their direction and speed of movement, both vertically and horizontally. As the use of radar for detecting meteorological phenomena is dealt with at some length in a report of a working group of the World Meteorological Organization[138] only a brief account of the subject is given in the following paragraphs.

Experiments have confirmed that the amount of back-scatter, and therefore the intensity of the radar echo, from spherical drops is in accordance with the theory originally propounded by Ryde[139]. The scattering function S, in the direction of the source of energy, of a spherical particle the diameter D of which is small compared with the wavelength λ of the radar transmission is given by:

$$S = \frac{\pi^5 D^6 |K|^2}{\lambda^4},$$

where $|K|$ is related to ϵ, the permittivity (or complex dielectric constant) of the material of the particle, by $|K| = (\epsilon - 1)/(\epsilon + 2)$. In a cloud the particles are not all of the same diameter. If N_D is the number per unit volume having diameter D the echo power is proportional to $\Sigma N_D S$, where

$$\Sigma N_D S = \frac{\pi^5 |K|^2}{\lambda^4} \Sigma N_D D^6 .$$

Suppose now a cloud of such particles completely intercepts at a range R a beam of radar pulses, as in A of Fig. 47, and that θ is the beam width in radians

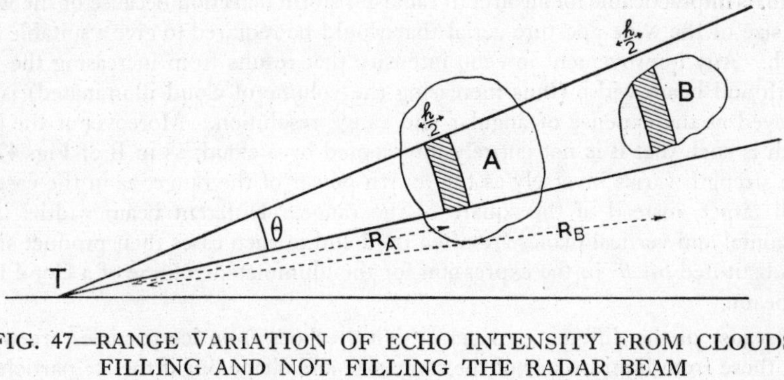

FIG. 47—RANGE VARIATION OF ECHO INTENSITY FROM CLOUDS FILLING AND NOT FILLING THE RADAR BEAM

Intensity from A, $1/R_A^2$; intensity from B, $1/R_B^4$

and h is the pulse length in space (i.e. the product of the pulse duration and the velocity of propagation), then the volume V of the cloud effectively " illuminated " by, and returning, a pulse is:

$$V = \frac{\pi R^2 \theta^2 h}{8},$$

the total energy arriving back at the radar at any given moment being that scattered by particles distributed over a depth $h/2$. The volume illuminated therefore has a total echo cross-section σ given by:

$$\sigma = \frac{\pi R^2 \theta^2 h \Sigma N_D S}{8}.$$

If the power and aerial gain of the radar transmitter are P_t and G_t the power delivered into the receiver from a target of echo cross-section σ is, as explained in Section 4.1.2, given by:

$$P_r = \frac{\sigma A_e P_t G_t}{16 \pi^2 R^4},$$

where A_e is the effective aperture of the receiving aerial. For a radar set using the same aerial for transmitting and receiving $G_t = 4\pi A_e/\lambda^2$, therefore:

$$P_r = \frac{\sigma A_e^2 P_t}{4\pi \lambda^2 R^4}.$$

The received echo power from a cloud intercepting the entire beam may then be expressed as:

$$P_r = \frac{\pi^5 A_e^2 \theta^2 K^2 h P_t \Sigma N_D D^6}{32 \lambda^6 R^2},$$

or, since θ^2 is proportional to λ^2/A_e,

$$P_r \propto \frac{A_e h P_t \Sigma N_D D^6}{\lambda^4 R^2}.$$

The echo intensity therefore gives a picture of the distribution of ND^6 and it is clear that drop size is a dominating factor governing the echo intensity and also that the shorter the wavelength the smaller the drops that can be detected. To obtain strong echoes at long ranges it is necessary to have high peak power, a wide-aperture aerial and long pulse length. There are, however, limitations to some of these factors. The effect of attenuation by rain, which has not been taken into account in the foregoing formulae, increases rapidly, with a consequent reduction in radar range, at wavelengths less than 3 cm. On the other hand a wavelength of 10 cm. is impracticable for an aircraft radar for storm detection because of the weight and size of the wide-aperture aerial that would be required to give a suitable beam width. Any improvement in echo intensity that results from increasing the pulse length and beam width (thus increasing the volume of cloud illuminated) is only achieved at the expense of angular and range resolution. Moreover if the beam width is such that it is not entirely intercepted by a cloud, as in B of Fig. 47, the echo strength varies inversely as the fourth power of the range, as in the case of a small target, instead of the square of the range. Different beam widths in the horizontal and vertical planes are often used and in such cases their product should be substituted for θ^2 in the expression for the illuminated volume of a cloud filling the beam.

Because of the difference in permittivity, echoes from ice particles are weaker than those from liquid water of the same size and shape but if the ice particles are wet they are as effective as liquid water.

11.5.2. Chief features of weather radar equipment

We may distinguish between two main requirements, namely, one for equipment for research into cloud and precipitation phenomena and the other for routine storm-warning purposes. For the former it is desirable to have equipment operating on a variety of wavelengths and with various forms of presentation. For the routine requirement, in which the primary aim is to locate rain clouds and showers and observe their movement and development, a less versatile equipment should suffice; the main consideration is that the apparatus should be capable of consistent performance for long periods without needing expert attention. Weather radar equipment differs from that used for upper wind measurement mainly in using longer pulse lengths and in having different aerial and scanning arrangements and types of display.

The power available in 10 cm. radars is such that it is extremely unlikely that a useful echo will be received from cloud particles, the diameter of which seldom exceeds 20 μ. Most of the weather echoes at this wavelength are caused by precipitation elements of about 1 mm. or more in diameter; attenuation by rain is negligible and does not restrict range performance appreciably. At 3 cm. also most of the echoes are due to precipitation elements, but attenuation by rain may be large enough to limit the range and if, for example, the beam passes through several showers, the more distant ones may be obscured by the nearer ones. All other factors being equal, therefore, heavy precipitation is detected more readily by using a wavelength of 10 cm., while for light precipitation it is better to use 3 cm. It is sometimes considered that a wavelength between 5 and 6 cm. represents the best compromise and it is certainly a more acceptable alternative than 10 cm. for airborne radar for which the size of a 10 cm. aerial is prohibitive.

At wavelengths of less than 1 cm. some, if not most, cloud particles are effective in producing echoes but attenuation by atmospheric gases, including water vapour, is important and limits the range of detection. There are absorption bands for water vapour at 13·5 and 1·6 mm. and for oxygen at 5·0 and 2·5 mm. Minimum attenuation occurs at about 8·6 mm. and with the transmitter powers that are available ranges up to about 15 km. should be obtainable at this wavelength. Radar equipment with such a wavelength is therefore suitable for the study of clouds—non-precipitating as well as precipitating types—provided the beam can be directed vertically or along high angles of elevation. With a vertical beam it is possible to determine the vertical distribution of cloud layers up to high levels.

It is desirable that the peak power output of the radar transmitter should be as high as possible. Useful ranges are obtainable with powers of about 500 kW. on the 10 cm. wavelength and 100 kW. on 3 cm. A high-gain aerial, with a beam width (measured at half-power points) of 2° or less is desirable, and the pulse length should be at least 2 μ sec.

A recent development of radar which uses the Doppler shift of frequency from moving targets has provided a useful tool for investigating the movement of precipitation particles in the direction of the beam.

The types of cathode ray tube display of echoes on weather radar sets are the A–scope, the plan-position indicator (P.P.I.) and the range–height indicator (R.H.I.), though these are not necessarily all provided on the same equipment. The A–scope indicates the echo as a " pip " or deflexion on a horizontal trace, the distance of the deflexion from the origin of the trace being proportional to the range of the source of the echo and the amplitude of the deflexion indicating the intensity of the echo. On the P.P.I. display, range and azimuth are indicated in polar co-ordinates. This necessitates scanning of the radar aerial round a vertical axis. The trace starts at the centre of the oscilloscope screen and follows a radius that is continuously locked to the direction of the beam. An echo is represented by a brightening of the trace at the appropriate range and azimuth. Most P.P.I. displays have limiting devices which cause echoes of high intensity to appear no stronger than weak echoes. The R.H.I. type of display is especially suitable for studying the vertical development of echoes. It is used in conjunction with an aerial that scans in the vertical plane from the horizon to a high angle of elevation at a fixed azimuth. The echoes are indicated on a screen displaying range and elevation in polar co-ordinates. As the maximum heights of the echoes are generally considerably less than the maximum horizontal range it is usual to expand the vertical scale relative to the horizontal scale on the screen.

A useful addition to the P.P.I. display is an off-centring device which allows the time-base origin to be displaced in any direction from the centre of the screen up to about one diameter of the tube. This enables a selected sector of the sweep to be expanded to cover the whole screen.

To facilitate the study of the echoes received on weather radar equipment it is very desirable that provision should be made for photographic equipment to record the traces on the display screens. In addition to manually operated single-shot cameras for use with all the tubes a cine-camera fitted with a time-lapse mechanism for automatically photographing the P.P.I. screen at predetermined intervals is a valuable aid.

11.5.3. Types of weather radar

Only a few radar equipments have been specially designed for cloud and precipitation observations and most of them operate on the 3 cm. wavelength. They include Decca and Marconi weather radars and one designed for the United States Air Force known as AN/CPS–9. These are primarily intended for routine operational purposes, and the first two are so used in the Meteorological Office. For research purposes in the Meteorological Office military radars have been used, one being a British 10 cm. set known as A.M.E.S. Type 21 and the other a 3 cm. American equipment AN/TPS–10, but recent developments include the use of 3 cm. Doppler radar and 8 mm. radar. Brief descriptions of some of these radars follow.

Radar A.M.E.S. Type 21.—This 10 cm. radar comprised two separate equipments, each with a nominal peak power of 500 kW. and a pulse duration of 1·9 μ sec. One was designed for plan-position display and had beam widths (to half-power points) of 1·25° and 6° in the horizontal and vertical planes respectively, while the other was for range–height display and had corresponding beam widths of 7·5° and 1·5°. In the P.P.I. equipment the aerial rotated four times per minute, usually with the axis of the beam at an elevation of 2°. Alternative maximum ranges of 30, 60 and 120 miles could be displayed on the 12 in. diameter P.P.I. screen. A second aerial with a wider beam in the vertical plane was mounted above the main aerial; it was used for co-operation with aircraft at high angles of elevation. In the R.H.I. equipment the aerial swept continuously in elevation between − 1° and + 20° once in 20 sec., the azimuth being manually controlled from the display position. An A–scope was provided as an alternative to the range–height display.

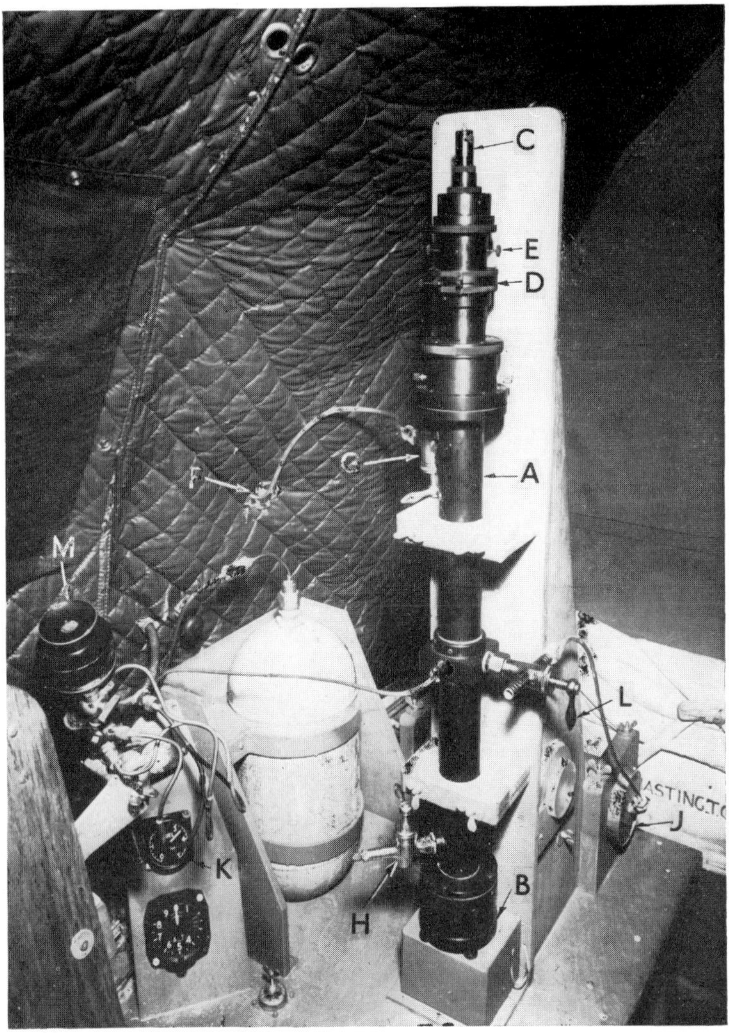

PHOTO-ELECTRIC NUCLEUS COUNTER

A, fog tube lined with moist blotting paper and sealed at the ends by electrically heated glass plates
B, photo-electric cell housing
C, lamp housing
D, iris diaphragm for lens mounted at upper end of fog tube (to produce parallel light beam in fog tube)
E, shutter control; to cut off light when determining dark current of photo-cell
F, pipe leading to intake mounted 10 cm. above aircraft skin and 1 m. ahead of propeller disc
G, H, manually operated valves to isolate sample after flushing of fog tube
J, filter system comprising cotton wool backing filter and Milipore fine filter to provide clean air for pressurization of sample
K, pressure indicator connected to fog tube
L, two-way manually operated valve used to admit pressurizing air to fog tube, seal fog tube and finally to permit sample to leak to the environment
M, Aitken nucleus counter

Plate XLIX

SCANNER UNIT OF DECCA STORM WARNING RADAR, TYPE 41

By courtesy of Marconi's Wireless Telegraph Co. Ltd.

SCANNER UNIT OF MARCONI STORM WARNING RADAR, TYPE SNW 51

Plate LI

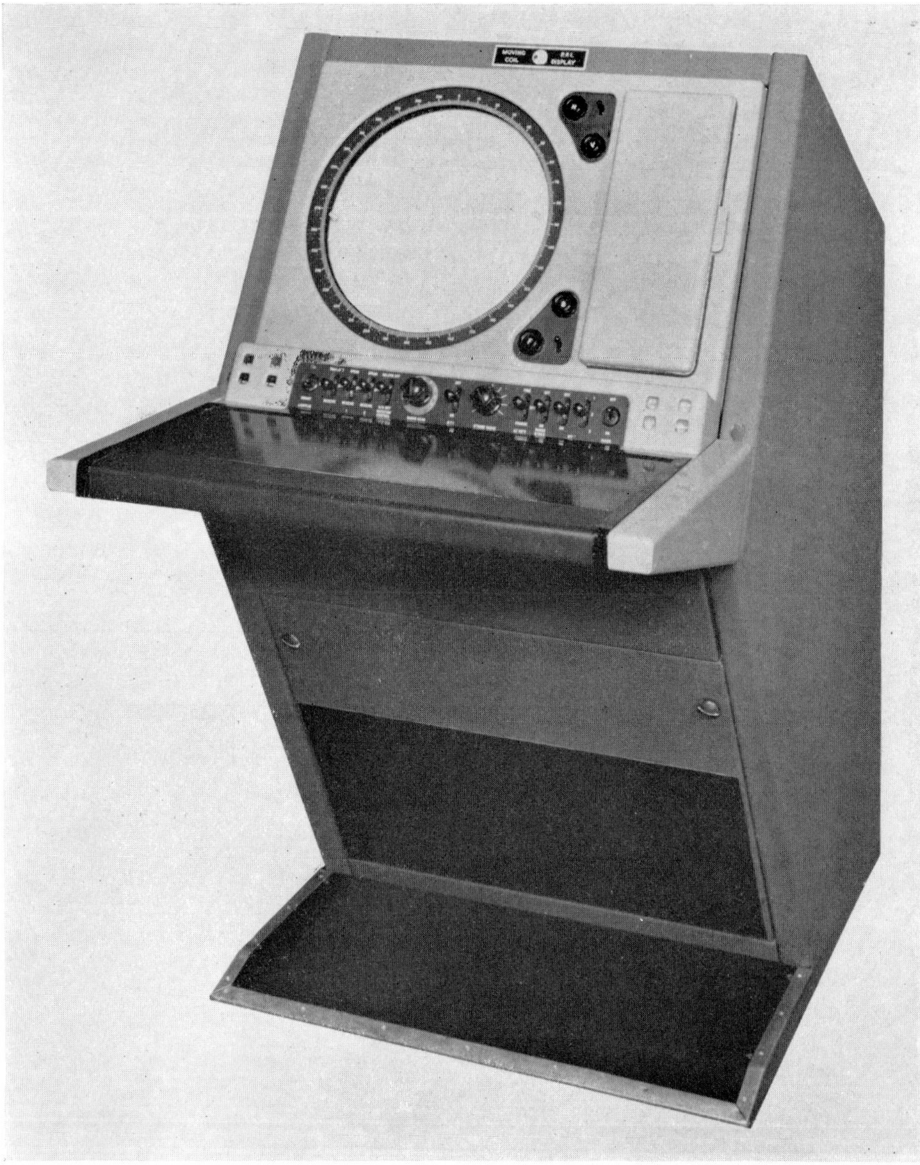

By courtesy of Marconi's Wireless Telegraph Co. Ltd.

DISPLAY UNIT OF MARCONI STORM WARNING RADAR, TYPE SNW 51

Decca storm-detection radar.—The Type 41 Mark 2 equipment transmits in the 3 cm. band with a nominal peak power of 20 kW. and alternative pulse durations of 2 and 0·2 μ sec. There are five main units comprising a scanner, with a rotating aerial and a radio-frequency unit, a receiver, display unit and power supply. The high-gain aerial, illustrated in Plate XLIX, has a 14 ft. horizontal-aperture reflector with a double curvature giving beam widths of 0·6° and 2·8° in the horizontal and vertical planes. It rotates at about 6 revolutions per minute and can be controlled in elevation between − 2° and + 28°. The receiver unit incorporates amplifiers, the pulse generator for the modulator, and power packs for the R.F. and display units. The latter has a 12 in. diameter P.P.I. tube on which time-bases giving range scales of various intervals up to 250 n. miles are provided. Angular accuracy is within 1°, while the accuracy of the range marker is within 2 per cent. of the indicated range. The equipment can be fitted with a 16 mm. cine-camera which can be triggered electronically to record every second, fourth, sixth or twelfth sweep of the P.P.I. trace.

Marconi storm-detection radar.—This 3 cm. equipment, Type SNW–51, is basically similar to the Decca set, but has a peak power output of 50 kW. It consists of four main units, namely, a scanner, a transmitter unit, display and power supply units. The scanner pedestal, shown in Plate L, carries an aerial reflector 13·5 ft. long horizontally and of parabolic section in the vertical plane, giving beam widths of 0·5° and 5° in these planes. The aerial can be rotated at rates of 5, 10, 15 or 20 r.p.m. and tilted through a range of elevation angles from − 2° to + 12°. Most of the electronic equipment, including the R.F. unit, the modulator and the receiver, is housed in the transmitter unit. The pulse duration is 2 μ sec. and the display is a 12 in. diameter P.P.I. (see Plate LI) provided with off-centring mechanism allowing the time-base origin to be displaced in any direction up to one radius of the tube. Time-base ranges of 40, 100, 150 and 200 n. miles are available and the accuracy is about 2 per cent. of the maximum range in use. Azimuth accuracy is 1°.

AN/CPS-9 storm-detection radar.—This set, which is of unusually high power for a 3 cm. radar, was designed at the United States Signal Corps Laboratories, and has been described by Johnson[140]. The separate units of which the set is comprised are an aerial pedestal, modulator, pulse generator, aerial control unit, presentation unit and a remote P.P.I. The peak transmitter power of 250 kW. and a conical beam of 1° width permit discrimination of weather echoes to very long ranges; the range display sweeps include one extending to 400 miles. The aerial can scan continuously in azimuth for P.P.I. display, or in sectors in either azimuth or elevation for R.H.I. display. Alternative pulse lengths of 0·5 and 5·0 μ sec. are used. The set is provided with complete calibration equipment, thus enabling echo intensities to be measured. It also has a separate range-amplitude display and an off-centre P.P.I. permitting an enlarged display of a limited area. With all these facilities available this equipment is well suited for research use as well as for routine observations.

Doppler radar.—The use of a 3 cm. pulsed Doppler radar for measuring the speed of fall of precipitation particles and, under certain conditions, for obtaining their size distribution has been described by Probert-Jones[141]. This equipment enables the radial velocities of targets to be determined from the Doppler frequency shift of the echo. It has a peak power output of 10 kW., a pulse width of 0·8 μ sec. and beam widths, to half power, of 2·5° by 3·5°. The information is produced in discrete form and is displayed as a matrix, each row corresponding to a range interval of about 150 m. and each column to a velocity interval of 0·5 or 1 m./sec. The echo power in each interval of range and velocity is summed and presented on the matrix as intensity modulation and the display is recorded by a 35 mm. camera. In its normal use, with a vertical beam, the radar gives the vertical velocities of the precipitation particles. It can also be used for wind measurements during precipitation by directing the beam along low angles of elevation to measure the component, along the beam, of the velocity of the particles which would be moving with the wind.

BIBLIOGRAPHY

1. London, British Standards Institution; Glossary of terms used in telecommunication. British Standard 204, London, 1943.
2. London, British Standards Institution; Glossary of terms used in automatic controlling and regulating systems. British Standard 1523, London, 1949 and 1954.
3. RONALDS, F.; Experiments made at the Kew Observatory on a new kite apparatus for meteorological observations. *Phil. Mag., London*, **31**, 1847, p. 191.
4. DINES, W. H.; A new meteorograph for kites. *Symons's met. Mag., London*, **39**, 1904, p. 109.
5. CAVE, C. J. P.; The great days of kite flying. *Weather, London*, **2**, 1947, p. 134.
6. WELSH, J.; An account of meteorological observations in four balloon ascents made under the direction of the Kew Observatory Committee of the British Association. *Phil. Trans., London*, 1853, p. 311 and *Ann. Chim. (Phys.), Paris*, **41**, 1854, p. 503.
7. DINES, W. H.; Two new light meteorographs for use with unmanned balloons. *Symons's met. Mag., London*, **41**, 1906, p. 101.
8. DINES, L. H. G.; The Dines balloon meteorograph and method of using it. London, 1929.
9. DINES, L. H. G.; A form of apparatus for obtaining samples of the atmosphere from great heights. *Quart. J. R. met. Soc., London*, **62**, 1936, p. 379.
10. SIMPSON, G. and SCRASE, F. J.; The distribution of electricity in thunderclouds. *Proc. roy. Soc., London*, A, **161**, 1937, p. 309.
11. SCHREIBER, P.; Bestimmung der Bewegung eines Luftballons durch trigonometrische Messungen von zwei Standpunkte. *Met. Z., Braunschweig*, **3**, 1886, p. 341.
12. KREMSER, V.; Pilotballons zum Studium der atmosphärischen Strömungen. *Met. Z., Braunschweig*, **10**, 1893, p. 198.
13. QUERVAIN, A. DE; Über die Bestimmung atmosphärischen Strömungen durch Registrier- und Pilotballons. *Met. Z., Braunschweig*, **23**, 1906, p. 149.
14. DINES, J. S.; On a theodolite with gear for recording the altitude and azimuth of pilot balloons. *Rep. Memor. adv. Comm. Aero., London*, No. 28, 2nd report on wind structure, 1910.
15. CAVE, C. J. P.; The structure of the atmosphere in clear weather. Cambridge, 1912.
16. DOUGLAS, C. K. M.; Weather observations from an aeroplane. *J. Scot. met. Soc., Edinburgh*, **17**, 1916, p. 65.
17. GLAZEBROOK, R.; Dictionary of applied physics. London, Vol. 3, 1923, p. 516.
18. DOBSON, G. M. B., BREWER, A. W. and CWILONG, B. M.; Meteorology of the lower stratosphere. *Proc. roy. Soc., London*, A, **185**, 1946, p. 144.
19. BAUMHAUER, E. H. von; Sur un météorographe universel destiné aux observations solitaires. *Arch. néerl. Sci., Haarlem*, **9**, 1874, p. 230.
20. WENSTROM, W. H.; Radiometeorography as applied to unmanned balloons. *Mon. Weath. Rev., Washington D.C.*, **62**, 1934, p. 221.
21. BLAIR, W. R. and LEWIS, H. M.; Radio tracking of meteorological balloons. *Proc. Inst. Radio Engrs., New York*, **19**, 1931, p. 1531.
22. IDRAC, P. and BUREAU, R.; Expériences sur la propagation des sondes radiotélègraphiques en altitude. *C.R. Acad. Sci., Paris*, **184**, 1827, p. 691.
23. MOLCHANOV, P.; Zur Technik der Erforschung den Atmosphäre. *Beitr. Phys. frei. Atmos., Leipzig*, **14**, 1928, p. 45.
24. DUCKERT, P.; Die Entwicklung der Telemeteorographie und ihrer Instrumentarien. *Beitr. Phys. frei. Atmos., Leipzig*, **18**, 1931, p. 68.
25. DUCKERT, P.; Das Radiosondenmodell Telefunken und seine Anwendung. *Beitr. Phys. frei. Atmos., Leipzig*, **20**, 1933, p. 303.
26. Berlin, Internationale Meteorologische Organisation, Internationale Aerologische Kommission; Über Radiosonde-Konstructionen. Berlin, 1937. Supplement, 1939.
27. THOMAS, H. A.; The determination of meteorological conditions by the use of radio-sounding balloons. *Proc. roy. Soc., London*, A, **167**, 1938, p. 227.
28. DYMOND, E. G.; The Kew radio sonde. *Proc. phys. Soc., London*, **59**, 1947, p. 645.
29. CORRIEZ, M. and PERLAT, A.; La méthode radiogoniométrique de l'Office National Météorologique pour la mésure de la direction et de la vitesse du vent par temps couvert. *Météorologie, Paris*, **11**, 1935, p. 368.

30. SMITH-ROSE, R. L. and HOPKINS, H. G.; The application of ultra-short-wave direction finding to radio sounding balloons. *Proc. phys. Soc., London*, **58,** 1946, p. 184.

31. HARRISON, D. N.; Direction-finding and the measurement of wind by radio. *Met. Mag., London*, **76,** 1947, p. 217.

32. CAVE, C. J. P. and WATSON WATT, R. A.; The study of radiotelegraphic atmospherics in relation to meteorology. *Quart. J. R. met. Soc., London*, **49,** 1923, p. 35.

33. WATSON WATT, R. A. and HERD, J. F.; An instantaneous direct-reading radiogoniometer. *J. Instn. elect. Engrs., London*, **64,** 1926, p. 611.

34. DIAMOND, H., HINMAN, W. S. and DUNMORE, F. W.; The development of a radio-meteorograph system for the Navy Department. *Bull. Amer. met. Soc., Milton, Mass.*, **18,** 1937, p. 73.

35. THOMAS, H. A.; A method of determining the position of an elevated radio transmitter. British Patent 28921/37.

36. KIRKMAN, R. A. and LEBEDDA, J. M.; Meteorological radio direction finding for measurement of upper winds. *J. Met., Milton, Mass.*, **5,** 1948, p. 28.

37. JONES, F. E., HOOPER, J. E. N. and ALDER, N. L.; The radar-sonde system for the measurement of upper wind and air data. *Proc. Instn. elect. Engrs., London*, **98,** Part II, 1951, p. 461.

38. London, Meteorological Office; Measurement of upper winds by means of pilot balloons. London, 1959.

39. SCHOUTE, C.; Ein Registrier-theodolit für Pilot-ballone. *Meded. ned, met. Inst., Utrecht*, No. 26, 1922.

40. SCULTETUS, H. R.; Erfahrungen mit dem Zeisschen vereinfachten Theodoliten für graphische Ballonaufnahmen nach Dr. Schoute. *Met. Z., Braunschweig*, **50,** 1933, p. 140.

41. CRANNA, R.; The new pilot balloon theodolite of the Meteorological Office. *Quart. J. R. met. Soc., London*, **64,** 1938, p. 613.

42. WHIPPLE, F. J. W.; The pilot balloon slide rule. *J. sci. Instrum., London*, **5,** 1928, p. 345.

43. POLLAK, L. W.; A graphical method of evaluating pilot-balloon ascents. *Bull. Amer. met. Soc., Milton, Mass.*, **22,** 1941, p. 345.

44. SAXTON, J. A.; The influence of atmospheric conditions on radar performance. *J. Inst. Navig., London*, **11,** 1958, p. 290.

45. HOOPER, A. H. and TAYLOR, A. P.; Radio refraction in the free atmosphere. *Met. Res. Pap., London*, No. 1021, 1956.

46. London, Meteorological Office; Handbook of radio-sounding technique. (In the press.)

47. ALLWOOD, H. I. S., BARTLETT, J. G. and DAVIES, G. T.; The measurement of the zero error of range in radar equipments. *Proc. Instn. elect. Engrs., London*, Part III A, **93,** 1946, p. 513.

48. BANNON, J. K.; Errors in winds measured with G.L. III radar equipment. *Met. Res. Pap., London*, No. 406, 1948.

49. HARRISON, D. N.; The use of the A.A. No. 3 Mark 7 radar for wind measurement. *Met. Res. Pap., London*, No. 672, 1951.

50. ELSE, C. V.; Wind-finding trials of radar A.A. No. 3 Mark 7 (second series). *Met. Res. Pap., London*, No. 922, 1955.

51. HARRISON, D. N.; Radar-wind comparisons on Ocean Weather Ships (second series). *Met. Res. Pap., London*, No. 928, 1955.

52. HALL, D. M. and BARKER, P. H.; The design of an improved primary radar, Type M.E.7, for radar wind observations. *Rep. Dep. sci. industr. Res. N.Z., Wellington*, No. R.244, 1955.

53. EWING, R. A.; The use of radar as an economical means of obtaining upper wind data. *Tech. Notes, N.Z. met. Serv., Wellington*, No. 111, 1954.

54. JONES, F. E., HOOPER, J. E. N. and ALDER, N. L.; The radar-sonde system for the measurement of upper wind and air data. *Proc. Instn. elect. Engrs., London*, Part II, **98,** 1951, p. 461.

55. GODDARD, N. E. and DELL, H. A.; A radar-sonde system for upper air measurement. *Philips tech. Rev., Eindhoven*, **19,** 1958, p. 258.

56. LUGEON, J.; La poste aérologique de la Station centrale de Météorologie à Payerne et les nouvelles méthodes suisse de radiosondage. *Ann. schweiz. met. ZentAnst., Zurich*, 1941, Nr. 7, p. 3.

57. MÜLLER, H. G.; Ein neues elektrisches Höhenwindmessverfahren. *Forsch.-u. Erfahr Ber. Reichsamt Wetterdienst, Berlin*, B, Nr. 20, 1944.

58. LUGEON, J.; Le radiosondage du vent par impulsion, sans répondeur. *C.R. Acad. Sci., Paris*, **236**, 1953, p. 2426.

59. L'HERMITTE, M.; Transpondeur radar pour la mésure du vent. *J. sci. Mét., Paris*, **6**, 1954, p. 123.

60. HOPKINS, H. G.; A loop direction-finder for ultra-short waves. *Wireless Engr., London*, **15**, 1938, p. 651.

61. FERENCE, M., JR.; Instruments and techniques for meteorological measurements. Compendium of meteorology, Boston, Mass., 1951, p. 1207.

62. TODD, W. and PETERSON, A.; A new high performance 1680 Mc./sec. rawinsonde system. Proceedings of the 5th Weather Radar Conference, Fort Monmouth, New Jersey, 1955, p. 207.

63. VÄISÄLÄ, V.; A new radiotheodolite. *Proc. Indian Acad. Sci., Bangalore*, **37A**, 1953, p. 223.

64. DU TOIT, P. S.; The Finnish radiosonde. News letter, Weather Bureau, Pretoria, No. 103, 1957, p. 8. (Unpublished.)

65. Geneva, World Meteorological Organization; Guide to international meteorological instrument and observing practice. *W.M.O. Publ. No.* 8, *Geneva*, 1954.

66. OCKENDEN, C. V.; Sferics. *Met. Mag., London*, **76**, 1947, p. 78.

67. HORNER, F.; A new design of radio direction-finder for locating thunderstorms. *Met. Mag., London*, **83**, 1954, p. 137.

68. ADCOCK, F. and CLARKE, C.; The location of thunderstorms by radio direction-finding. *J. Instn. elect. Engrs., London*, Part III, **94**, 1947, p. 118.

69. HORNER, F.; The accuracy of the location of sources of atmospherics by radio direction-finding. *J. Instn. elect. Engrs., London*, Part III, **101**, 1954, p. 383.

70. Geneva, World Meteorological Organization; An instrument for counting local lightning flashes. *W.M.O. Bulletin, Geneva*, **8**, 1959, p. 14.

71. CONOVER, W. C. and STROUD, W. G.; A high-altitude radio-sonde hypsometer. *J. Met., Milton, Mass.*, **15**, 1958, p. 63.

72. WEXLER, A.; Electric hygrometers. *Circ. nat. Bur. Stand., Washington, D.C.*, No. 586, 1957.

73. FRANKENBERGER, E.; Neue trägheitsarme Messelelemente für Haarhygrometer. *Forsch. -u. ErfahrBer. Reichsamt Wetterdienst, Berlin*, A, 1944, No. 26.

74. DUNMORE, F. N.; An improved electric hygrometer. *J. Res. nat. Bur. Stand. Washington, D.C.*, **23**, 1939, p. 701.

75. WEXLER, A., GARFINKEL, S. B., JONES, F. E., HASEGAWA, S. and KRINSKY, A.; A fast responding electric hygrometer. *J. Res. nat. Bur. Stand., Washington, D.C.*, **55**, 1955, p. 71.

76. SMITH, W. K. and HOEFLICH, N. J.; The carbon film electric hygrometer element. *Bull. Amer. met. Soc., Milton, Mass.*, **35**, 1954, p. 60.

77. TAYLOR, E. S.; New instrument for the moisture analysis of freon fluorinated hydrocarbons. *Refrig. engng., New York*, **64**, 1956, p. 41.

78. CRAWSHAW, J. K. and DAVIDSON, F. G.; Electric hygrometers for measurements in gases down to a few parts per million. *J. Sci. Instrum., London*, **36**, 1959, p. 121.

79. VÄISÄLÄ, V.; The Finnish radio-sound and its use. *Mitt. met. Inst. Univ. Helsingf., Helsinki*, No. 35, 1937.

80. Toronto, Meteorological Branch, Department of Transport; Manual 91, The Canadian radiosonde and rawinsonde system. *Circ. met. Div. Dep. Transp., Toronto*, No. 2877, 1957.

81. BUREAU, R.; Les radiosondages météorologiques. *Onde élect., Paris*, **14**, 1935, p. 10.

82. VENKITESHWARAN, S. P., THATTE, R. P. and KESHAVAMURTHY, A.; The fan type radio-meteorograph. *Sci. Notes met. Dep. India, Poona*, **9**, No. 113, 1948, p. 127.

83. LUGEON, J., ACKERMANN, P. and BOHNENBLUST, M.; La radiosonde suisse. *Ann. schweiz. met. ZentAnst., Zurich*, 1948.

84. KHAKHALIN, V. S.; Radiosondes. Leningrad, 1955. (In Russian.)

85. HINZPETER, H.; A new radiosonde. *Ann. Met., Hamburg*, **4**, 1951, p. 393.

86. ISONO, K. and HUZIWARA, Y.; Improved code sending type radiosonde. *Pap. Met. Geophys., Tokyo*, **1**, 1950, p. 299.

87. DIAMOND, H., HINMAN, W. S. and DUNMORE, F. W.; A method for the investigation of upper-air phenomena and its application to radio meteorography. *J. Res. nat. Bur. Stand., Washington, D.C.*, **20**, 1938, p. 369.

88. DIAMOND, H., HINMAN, W. S., DUNMORE, F. W. and LAPHAM, E. G.; An improved radio sonde and its performance. *J. Res. nat. Bur. Stand., Washington, D.C.*, **25**, 1940, p. 327.

89. HAUER, A. and van TOL, M.; A radiosonde for meteorological observations. *Philips tech. Rev., Eindhoven,* **16,** 1954, p. 148.

90. ROSSI, V.; A new Finnish radiosonde. *Mitt. met. Inst. Univ. Helsingf., Helsinki,* No. 48, 1957.

91. SITTEL, K. and MENZER, E.; An improved fully electric radiosonde. *Bull. Amer. met. Soc., Milton, Mass.,* **31,** 1950, p. 341.

92. SPILHAUS, A. F., SCHNEIDER, C. S. and MOORE, C. B.; Controlled-altitude free balloons. *J. Met., Milton, Mass.,* **5,** 1948, p. 130.

93. ANDERSON, A. D. and MASTENBROOK, H. J.; A new upper air data system—the transosonde. *Bull. Amer. met. Soc., Milton, Mass.,* **37,** 1956, p. 342.

94. BRAILSFORD, H. D.; A new code transmitting radiosonde. *J. Met., Milton, Mass.,* **6,** 1949, p. 360.

95. BELIN, R. E.; A radiosonde method for atmospheric potential gradient measurements. *Proc. phys. Soc., London,* **60,** 1948, p. 381.

96. CHAPMAN, S.; Thunderstorm electricity in relation to rain and snow particles. Thunderstorm electricity, Chicago, 1953, p. 207.

97. KOENIGSFELD, L. and PIRAUX, P.; Un nouvel électromètre portatif; son application à la radiosonde. *Mém. Inst. mét. Belg., Bruxelles,* No. 45, 1950.

98. KOENIGSFELD, L.; La mésure de la conductibilité par radiosonde. *Publ. Inst. mét. Belg., Bruxelles,* Series B, No. 18, 1955.

99. JONES, O. C., MADDEVER, R. S. and SANDERS, S. H.; Radio-sonde measurements of vertical electrical field and polar conductivity. *J. sci. Instrum., London,* **36,** 1959, p. 24.

100. VENKITESHWARAN, S. P., DHAR, N. C. and HUDDAR, B. B.; On the measurement of the electrical potential gradient in the upper air over Poona by radiosondes. *Proc. Indian Acad. Sci., Bangalore,* **37A,** 1953, p. 260.

101. VENKITESHWARAN, S. P., GUPTA, B. K. and HUDDAR, B. B.; Measurement of the electrical conductivity in the upper air by radiosonde. *Proc. Indian Acad. Sci., Bangalore,* **38A,** 1954, p. 109.

102. CORONITI, S. C., NAZAREK, A., STERGIS, C. G., KOTAS, D. E., SEYMOUR, D. W. and WERME, J. V.; Balloon borne conductivity meter. *Instrumentation geophys. Res., Cambridge, Mass.,* No. 3, 1954.

103. LUGEON, J. and BOHNENBLUST, M.; Radiosondage du gradient de potentiel et de la conductabilité électrique de l'air. *Ann. schweiz. met. ZentAnst., Zurich,* **93,** 1956, Nr. 6, p. 49.

104. COBLENTZ, W. W. and STAIR, R.; Radiometric measurement of ultraviolet solar radiation intensities in the stratosphere. *J. Res. nat. Bur. Stand., Washington, D.C.,* **20,** 1938, p. 185.

105. KULCKE, W. and PAETZOLD, H.-K.; Über eine Radiosonde zur Bestimmung den vertikalen Ozonverteilung. *Ann. Met., Hamburg,* **8,** 1957, p. 47.

106. SCRASE, F. J.; Radiation and lag errors of the Meteorological Office radiosonde. *Quart. J. R. met. Soc., London,* **80,** 1954, p. 565.

107. SCRASE, F. J.; The application of radiation and lag corrections to temperature measured with the Meteorological Office radio-sonde. *Met. Mag., London,* **85,** 1956, p. 65.

108. GLÜCKAUF, E.; Investigations on absorption hygrometers at low temperatures. *Proc. phys. Soc., London,* **59,** 1947, p. 344.

109. Zurich, World Meteorological Organization, Commission for Instruments and Methods of Observation, Working Group on Comparison of Radiosondes; Final report of the meeting in Zurich (Switzerland). Zurich, 1956.

110. GOLDSTEIN, S.; Modern developments in fluid mechanics. Vol. II. Oxford, 1943.

111. FROST, R.; The rate of ascent of free balloons. *Met. Mag., London,* **88,** 1959, p. 70.

112. London, Meteorological Office; Meteorological air observer's handbook. London, 1945.

113. Montreal, International Civil Aviation Organization; Manual of ICAO standard atmosphere. Document No. 7488, Montreal, 1954.

114. HOATHER, W. M.; The physical principles of some basic aircraft instruments. *J. sci. Instrum., London,* **25,** 1948, p. 113.

115. BULL, G. A.; Notes on the I.C.A.N. altimeter and height and air speed computer. *Met. Mag., London,* **77,** 1948, pp. 6, 29.

116. CLARK, D. D.; A thermometer for high-speed aircraft. *J. sci. Instrum., London,* **35,** 1958, p. 433.

117. TERADA, K. and JAMAMOTO, G.; Methods for measuring air temperature on a high-speed airplane. *J. Met., Milton, Mass.,* **4,** 1947, p. 201.

118. VONNEGUT, B.; A vortex thermometer for measuring true air temperature and true air speeds in flight. *Rev. sci. Instrum., New York,* **21,** 1950, p. 136.

119. HINKEL, C. H.; Report on flight tests of a prototype conical-head aircraft thermometer. *Met. Res. Pap., London,* No. 1043, 1957.

120. CLARK, D. D.; An assessment of the probable causes of variation of the speed-correction coefficient of aircraft thermometers. *Tech. Rep. aero. Res. Comm., London,* No. 3127, 1959.

121. DOBSON, G. M. B., BREWER, A. W. and CWILONG, B. M.; Meteorology of the lower stratosphere. *Proc. roy. Soc., London,* A, **185,** 1946, p. 144.

122. BREWER, A. W., CWILONG, B. M. and DOBSON, G. M. B.; Measurement of absolute humidity in extremely dry air. *Proc. phys. Soc., London,* **60,** 1948, p. 52.

123. GOLDSMITH, P.; A method of increasing the range of the Dobson–Brewer frost-point hygrometer in jet aircraft. *Quart. J. R. met. Soc., London,* **81,** 1955, p. 607.

124. THORNTHWAITE, C. W. and OWEN, J. C.; A dew-point recorder for measuring atmospheric moisture. *Mon. Weath. Rev., Washington, D.C.,* **68,** 1940, p. 315.

125. WOOD, R. C., FOSKETT, L. W. and BOSTER, N. B. [error for FOSTER, N. B.—*Ed.*]; The infra-red absorption hygrometer. A progress report. *Proc. instrum. Soc. Amer., Pittsburgh, Pa.,* **9,** Part 1, 1954, No. 54–36–1.

126. WOOD, R. C.; Improved infra-red absorption spectra hygrometer. *Rev. sci. Instrum., New York,* **29,** 1958, p. 36.

127. MURGATROYD, R. J. and HELLIWELL, N. C.; The measurement of wind from aircraft using a Doppler navigation system. *J. Inst. Navig., London,* **12,** 1959, p. 84.

128. WARNER, J. and NEWNHAM, T. D.; A new method of measurement of cloud water content. *Quart. J. R. met. Soc., London,* **78,** 1952, p. 46.

129. DAY, G. J.; A refrigerated disc icing meter. *Met. Res. Pap., London,* No. 916, 1955.

130. DURBIN, W. G.; Recent aircraft research into cloud physics. *Weather, London,* **13,** 1958, p.143.

131. SQUIRES, P. and GILLESPIE, C. A.; A cloud droplet sampler for use on aircraft. *Quart. J. R. met. Soc., London,* **78,** 1952, p. 387.

132. GARROD, M. P.; Recent developments in the measurement of precipitation elements from aircraft. *Met. Res. Pap., London,* No. 1050, 1957.

133. SCHAEFER, V. J.; The formation of ice crystals in the laboratory and the atmosphere. *Chem. Rev., Baltimore,* **44,** 1949, p. 291.

134. NOLAN, P. J. and POLLAK, L. W.; The calibration of a photo-electric nucleus counter. *Proc. R. Irish Acad., Dublin,* A, **51,** 1946, p. 9.

135. VITTORI, O.; Sur la détermination de la distribution des diamètres des particules atmosphériques de chlorure avec la réaction de Liesegang. *Geofis. pur. appl., Milan,* **31,** 1955, p. 90.

136. SMITH, E. N. and HEFFERNAN, K. J.; Airborne measurements of the concentration of natural and artificial freezing nuclei. *Quart. J. R. met. Soc., London,* **80,** 1954, p. 182.

137. WARNER, J.; An instrument for measuring freezing nucleus concentration. *Bull. Obs. Puy de Dôme, Clermont Ferrand,* No. 2, 1957, p. 33.

138. Geneva, World Meteorological Organization; Use of ground-based radar in meteorology (excluding upper wind measurements). *W.M.O. Tech. Notes, Geneva,* No. 27, 1959.

139. RYDE, J. W.; The attenuation and radar echoes produced at centimetre wave-lengths by various meteorological phenomena. Meteorological factors in radio-wave propagation; Report of a Conference held on 8 April 1946 at The Royal Institution, London, by the Physical Society and the Royal Meteorological Society. London, 1946, p. 169.

140. JOHNSON, R. H.; New Air Force weather equipment. *Weatherwise, Milton, Mass.,* **7,** 1954, p. 82.

141. PROBERT-JONES, J. R.; The meteorological use of pulsed Doppler radar. *Nature, London,* **186,** 1960, p. 271.

142. BREWER, A. W. and MILFORD, J. R.; The Oxford–Kew ozone sonde. *Proc. roy. Soc., London,* A, **256,** 1960, p. 470.

APPENDIX

A BRIEF GLOSSARY OF TECHNICAL TERMS

A-SCOPE OR TYPE-A DISPLAY. See RANGE-AMPLITUDE DISPLAY.

AERIAL GAIN. The ratio of the power radiated in an elementary solid angle in a given direction to that which would be produced in the same solid angle if the same total power were radiated isotropically. Alternatively it is the ratio of the power which must be supplied to an isotropic radiator to produce a given field at a given point to that which must be supplied to the aerial to produce the same field at the same point. The gain in the direction of maximum radiation may be called the maximum gain. Since power densities are the same whether the aerial is transmitting or receiving, the power gain is the same for both functions.

APERTURE (of radar aerial). The effective area of the radiating surface in the plane normal to the principal direction of propagation.

ATMOSPHERICS (see also SFERIC). Electromagnetic radiation due to natural causes, and the disturbances produced thereby on a radio receiver.

AUDIO-FREQUENCY. A frequency of oscillation within the range audible to the normal human ear, namely, from about 30 to 8000 c./s.

AUTOMATIC FOLLOWING. A system in which a mechanism actuated by a radar echo automatically keeps the radar beam pointing at an object moving relatively to the radar set and automatically measures the range.

AUTOMATIC GAIN (OR VOLUME) CONTROL. A device to keep the output of an amplifier substantially constant despite variations in input level.

B-SCOPE OR TYPE-B DISPLAY. See RANGE–BEARING DISPLAY.

BEAM-SWITCHING. A method of determining the direction of a remote object by comparison of the signals corresponding to two or more successive beam angles, differing slightly from the direction of the object. The motion of the beam may be continuous and periodic or discontinuous. The method is frequently referred to as SPLIT.

BEAM WIDTH. The angular width of a beam of radiation, measured between the directions in which the power intensity is a specified fraction, usually one-half, of the maximum.

BEAT-FREQUENCY OSCILLATOR. An oscillation generator in which an output alternating voltage of approximately sinusoidal wave form is produced by the beating of two radio-frequency oscillations.

BLOCKING OSCILLATOR. A self-quenching oscillator which becomes cut off before half a cycle of oscillation is completed.

BUFFER STAGE. A valve-coupled stage specially designed to prevent load fluctuations of subsequent stages from affecting the frequency or other characteristics of preceding circuits.

CARRIER FREQUENCY. The frequency of a modulated wave before modulation is applied.

CATHODE-FOLLOWER CIRCUIT. A circuit in which the output load is included in the cathode circuit of a valve, and the input to the valve is applied between the grid and the remote end of the cathode load.

CATHODE RAY TUBE DISPLAY. The presentation of a received signal on the fluorescent screen of a cathode ray tube.

CERAUNOMETER. An instrument for counting the total number of lightning flashes in the vicinity of a station.

CORNER REFLECTOR. A reflector, consisting of three mutually perpendicular, intersecting, plane conducting surfaces, which reflects incident waves parallel to their direction of incidence.

DECIBEL. A unit which expresses the ratio between two levels of power. With P_1 and P_2 designating two amounts of power and N the number of decibels denoting their ratio,

$$N = 10 \log_{10} (P_1/P_2).$$

Provided the powers are dissipated in equal impedances the ratio may be expressed in terms of voltage or current, thus:

$$N = 2 \times 10 \log_{10} (V_1/V_2) = 2 \times 10 \log_{10} (C_1/C_2),$$

power being proportional to the square of voltage and current.

DEFLECTOR PLATES AND COILS. Parts of a cathode ray tube controlling the position of incidence of the electron beam on the screen, by means of an electrostatic field produced by a potential applied to the plates, and a magnetic field produced by a current through the coils.

DIPOLE. A straight aerial symmetrical in regard to its standing-wave current and usually approximately half a wavelength long.

DISH OR PARABOLOID. The paraboloid reflector of a radar aerial system. The aerial is placed at the focal point and the radio energy is beamed in the desired direction.

DISPLAY OR PRESENTATION. Visual presentation of received signals.

ECHO. In radar: (i) The radio-frequency energy received after reflection from an object. (ii) The effect of (i) on a radar display, e.g. by a deflexion or change of intensity of a cathode ray tube trace.

FEED-BACK. The return of energy from the output to the input of a thermionic valve or system of valves and associated circuits. Feed-back may be either positive or negative, i.e. tending to increase or to decrease the amplification, and may be made dependent on the voltage or the current of the output load or independent of the impedance of the load.

FEEDER. The non-radiating electrical conductor joining an aerial to its source of energy or to its receiver. Also referred to as a transmission line.

FILTER. A network designed to transmit currents of frequencies within one or more frequency bands and to attenuate currents of other frequencies. Filters having single transmission bands extending (*a*) from zero frequency to the cut-off frequency, (*b*) from the cut-off frequency to infinite frequency, and (*c*) from one cut-off frequency to another are called low-pass, high-pass and band-pass filters respectively.

FREQUENCY CHANGER OR MIXER. A device, associated with a beat oscillator, which delivers output at a frequency differing from the input frequency.

GAIN. The increase in power in transmission from one point to another.

GATING. Making a channel effective for a desired interval, beginning at an instant separated from a reference instant by a desired interval. On a radar range display a gate is a small portion of the trace; its position is variable and the response must fall within the gate in order to be indicated by the set.

HALF-WAVE DIPOLE. See DIPOLE.

INTERMEDIATE FREQUENCY. A frequency to which that of the incoming signal is changed in superheterodyne reception.

KLYSTRON. A valve in which the electron stream is modulated in velocity and the electrodes of the output circuit (and also possibly of the input circuit) are combined to form a resonant circuit of a special type known as a rhumbatron.

LOBE. The boundary of a volume inside which the power of a beam of radiation is everywhere greater than a chosen value. In practice there are side lobes associated with subsidiary beams. The lobe pattern as a whole is a diffraction phenomenon.

LOCKING PULSE. A pulse emitted by a radar transmitter or its associated circuits at the instant the pulse of radio energy is emitted, which is applied to the receiver where it is used to start a time-base. This ensures that the echo appears at the same position on the range display on successive sweeps of the time-base.

LOCK LINE. Cable over which the locking pulse is transmitted.

MAGNETIC AMPLIFIER. An electrical amplifier in which the amplification is produced by transductors.

MAGNETRON. A thermionic valve, the electron path of which is controlled by a magnetic field.

MAGSLIP. See SYNCHRO.

MIXER. See FREQUENCY CHANGER.

MODULATION. The process by which the amplitude, frequency or phase of a carrier wave is modified in accordance with the characteristics of a signal.

MODULATION FACTOR. The ratio of half the difference of the maximum and the minimum amplitude or frequency to the mean amplitude or frequency of a modulated wave.

NOISE. Unwanted energy (or the voltage produced), usually of random character, present in a transmission system, due to any causes. In particular applications the term may be limited to noise of specified origin.

PARABOLOID. See DISH.

PERMANENT ECHO (P.E.). A radar echo, at a fixed station, due to any fixed object, such as a tall mast. Such echoes are useful for checking the alignment of the radar set if the exact positions of the objects are known.

PHANTASTRON. A single valve relay designed to produce an output variation at some pre-determined instant after the application of a triggering signal.

PIP. A deflexion or change of intensity, on a cathode ray tube display, produced as a calibration or range marker.

PLAN-POSITION INDICATOR (P.P.I.). A radar display indicating, as on a map, the relative position of echo-producing objects within the range of the radar set. In general the centre of the P.P.I. display corresponds to the position of the radar set, but this is not the case with an off-centre plan display.

POLAR DIAGRAM. A graphical representation, using polar co-ordinates, of the variation of the power of the signal received from a fixed transmitter with the angular displacement of the moving part of the direction-finding system.

POLARIZATION, PLANE OF. The plane containing the directions of the electric force and of propagation in the case of plane-polarized waves.

PULSE. An e.m.f. of sharp-fronted (square) wave form and of short duration.

PULSE WIDTH. The duration of a pulse.

PUSH–PULL VALVE OPERATION. The use of a pair of similar valves, or of a double valve, to the control grids of which are applied voltages which are equal in amplitude and opposite in phase, the corresponding outputs being combined in a balanced-output circuit.

RADAR. The use of radio waves, reflected or automatically re-transmitted, to gain information concerning a distant object. In the case of automatic re-transmission, the delay in evoking the re-transmission must be sufficiently short and precise for the range to be measurable.

RADIO DIRECTION-FINDING. The determination of the direction of an object by means of its own independent emissions.

RADIO-FREQUENCY. Any frequency at which electromagnetic radiation is used for telecommunication.

RADIOGONIOMETER. An instrument which when coupled to a suitable fixed aerial system enables the bearing of arriving waves to be determined by rotation of a moving part.

RANGE-AMPLITUDE DISPLAY. A radar display in which a time-base provides the range scale on which echoes appear as deflexions normal to the base. The amplitudes of the deflexions depend on the intensities of the echoes. A range-amplitude display in which the time-base is sensibly a straight line is often referred to as a Type-A display or A-scope.

RANGE–BEARING DISPLAY. A radar display in which an echo appears on the screen as a bright spot whose rectangular co-ordinates indicate the range and bearing of the object. This is often referred to as a Type-B display or B-scope.

RANGE–HEIGHT INDICATOR (R.H.I.). A radar display which shows simultaneously angular elevation, slant range and height of targets in an elevation sector scanned by the beam. Height is indicated by straight horizontal marker lines and slant range by curved lines.

RANGE MARKER. A visual discontinuity in the time-base of a radar display for measuring the range or for calibrating the time-base. In a P.P.I. the range marker usually appears as a visible ring.

RESPONDER. See TRANSPONDER.

SCAN. To explore a region by the automatic continuous variation of the direction of a radar beam.

SELSYN. See SYNCHRO.

SERVO-SYSTEM. A monitored automatic control system which includes a power amplifier in the main forward path.

SERVO-MOTOR. The final control element in a servo-mechanism. It is the motor which receives the output from the amplifier element and which drives the load.

SFERIC. The electromagnetic wave resulting from a lightning discharge in the atmosphere, and its manifestation in equipment employed for meteorological purposes.

SIDE LOBE. See LOBE.

SIGNAL-TO-NOISE RATIO. The ratio of the effective voltage at the nominal signal level of the desired signal to that of noise.

SLANT RANGE. The straight-line distance from a radar set to a target. This is the range normally indicated by the range scale on the set.

SPLIT. See BEAM-SWITCHING.

STROBE, TO. To select a desired epoch of a recurrent phenomenon.

STROBE PULSE. A pulse, of duration less than the period of a recurrent phenomenon, used for scrutinizing a particular epoch of that phenomenon.

STROBE MARKER. A small bright spot, or other discontinuity, produced on the line trace of a radar display to indicate that part of the time-base which is receiving attention.

SUPERHETERODYNE RECEPTION. A method of beat reception in which one or more frequency changes take place before final detection. The first frequency change is to an ultra-sonic frequency.

SUPER-REGENERATIVE RECEPTION. A method of reception employing amplification in which feed-back is adjusted to a point at which oscillation occurs, or is liable to occur, any oscillation produced being periodically suppressed.

SWEEP. See TRACE.

SYNCHRO. A generic term for a class of electromechanical devices used for data transmission. Such devices are more often known by proprietary names such as MAGSLIP and SELSYN.

SYNCHRO TRANSMITTER. A synchro, the rotor of which is mechanically positioned for transmitting electrical information corresponding to the angular position of the rotor.

SYNCHRO RECEIVER. A synchro, the rotor of which is free to turn and which, when energized, develops a torque dependent on the difference between its synchro angle and the electrical angle received from its connected transmitter.

SYNCHRO RESOLVER. A synchro capable of resolving an electrical vector into two perpendicular components or of compounding, in conjunction with a servo-mechanism, two perpendicular vectors into a resultant vector.

SYNCHRONISING SIGNAL OR SYNC-PULSE. A signal sent out periodically by the transmitter in order to keep the receiving system in synchronism.

TARGET. An object from which a radar echo is received or sought.

THYRATRON. A gas-filled tube in which the passage of current in an ionized gas or vapour is influenced by the voltage applied through a control electrode.

TIME-BASE. The deflexion of the spot on a cathode ray tube which is defined in relation to time. In a linear time-base the spot moves at a constant speed in the direction of the time scale.

TIME-BASE GENERATOR. Apparatus for producing the necessary voltage or current for establishing a time-base.

TRACE OR SWEEP. The movement of the spot on a cathode ray tube as determined by the time-base.

TRANSDUCER. A device used for converting a signal or physical quantity of one kind into a corresponding physical quantity of another kind.

TRANSDUCTOR. A device consisting of one or more ferro-magnetic cores with windings, by means of which an a.c. voltage or current can be varied by an independent voltage or current, utilizing saturation phenomena in the magnetic circuit. Used in magnetic amplifiers.

TRANSMISSION LINE. See FEEDER.

TRANSPONDER. A unit which receives pulses from a radar set or interrogator and, in response to the received pulse, transmits a pulse or sequence of pulses which can be recognized by the interrogating station. Sometimes known as a RESPONDER.

VIDEO-FREQUENCIES. The frequencies of modulated signals which may be applied to a cathode ray tube to produce a radar display (or television picture). They may range from 100 c./s. to several megacycles per second.

WAVEGUIDE. An elongated volume of air or other dielectric, bounded along its length by one or more surfaces which may be conducting or may be surfaces of discontinuity of permittivity and/or permeability; used for guided transmission of electric waves. Usually restricted to a rectangular or circular metal tube of such dimensions that energy at the desired frequency can be propagated.

INDEX

(79967) Wt. 2259/B.M.4013 K 20 2/61 Hw. S.O. Code No.40-15-22*